# Study Guide

# Biology
## MILLER · LEVINE

**Prentice Hall**
Englewood Cliffs, New Jersey
Needham, Massachusetts

**BIOLOGY**
Miller • Levine

**Study Guide**

ISBN 0-13-803040-5

7 8 9 10     03 02 01 00 99

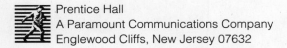

Prentice Hall
A Paramount Communications Company
Englewood Cliffs, New Jersey 07632

# Contents

# About the Study Guide

The Study Guide that accompanies *Biology* by Kenneth Miller and Joseph Levine has been specifically designed to help you in your biology course. The Study Guide concentrates on the key facts and concepts presented in each chapter of the textbook. Each chapter in the Study Guide is divided into the same numbered sections as in the textbook. By completing each Study Guide section after you have read the section in the textbook, you will be better able to understand and remember the important points made in the chapter. In addition, your ability to tie information together in order to see the "big picture" in biology will be greatly improved. A description of the organization of the Study Guide follows.

## Section Review

Each section in the Study Guide begins with a Section Review. This review material stresses the key concepts and facts you should focus on in that particular section. As you read the Section Review, try to relate the material you have read in the chapter to the material stressed in the review. If parts of the Section Review are not clear to you, go back to that part of the section in the textbook and read it again.

An alternate way of using the Section Review is to read it before you read the section in the textbook. In that way you will be alerted to the important facts the authors thought you should concentrate on. Used in this manner, the Section Review can be a prereading guide to the chapter material.

## Section Activities

Following the Section Review you will find specific workbook activities designed to help you read and understand the textbook. These activities vary in format and type, but many help in building vocabulary skills and finding the main ideas within a section. Still others will enable you to take the information in the chapter and apply it to a real-world situation.

## Concept Mapping

Once you have read the Section Review and completed the Section Activities, you will be ready to begin concept mapping that particular section of the chapter. A short note reminds you to turn to the partial concept map found on the last page of every chapter in the Study Guide.

## Using the Writing Process

The ability to write clearly and accurately is not only important to scientists, it is important to all of us in our everyday lives. Each chapter in the Study Guide includes a writing suggestion. Essays, short stories, television scripts, debate outlines, and advertising campaigns are among the many kinds of writing suggestions you will be given. Although these writing assignments do require you to read and understand the information presented in the textbook chapter, they have also been designed to be enjoyable and will allow you a great deal of creative flexibility.

## Concept Mapping

As noted earlier, you will be asked to complete a partial concept map for each chapter. It is suggested that you fill in the partial concept map as you finish each chapter section. However, you may want to do the concept map after completing the entire chapter. Please note that the first chapter includes a complete concept map. Chapters that follow include the partial concept map.

Concept maps will help you to graphically organize the important facts and concepts presented in the chapter and to tie them together. Unlike a simple outline that takes each topic and divides it into subtopics, the concept map allows you to show the relationships between different concepts. By completing the concept map, you will be better able not only to understand the important aspects of the chapter, but also to show how these concepts fit together into the big picture called biology.

Concept maps can be done in a variety of ways. There is no one way to do a concept map for a particular chapter. Your concept map will almost certainly be different from those of the other students. That is to be expected—just as your outline of a particular chapter would also be different. Some concept maps may be very detailed and others may stress only the largest concepts and their relationships. The way you do your concept map should be the way it best helps you read and understand the chapter. In fact, you may decide not to use the partial concept map already printed in each chapter but to begin your own from scratch. As long as your teacher agrees, that is a perfectly acceptable decision.

To demonstrate the different types of concept maps that can be completed for a chapter, examine the concept maps labeled Concept Map 1 and Concept Map 1A. Concept Map 1 is the completed concept map printed for Chapter 1 in this Study Guide. Concept Map 1A is a much more detailed version of a concept map for Chapter 1. Neither is wrong. One just shows more detail than the other.

**Concept Map 1**

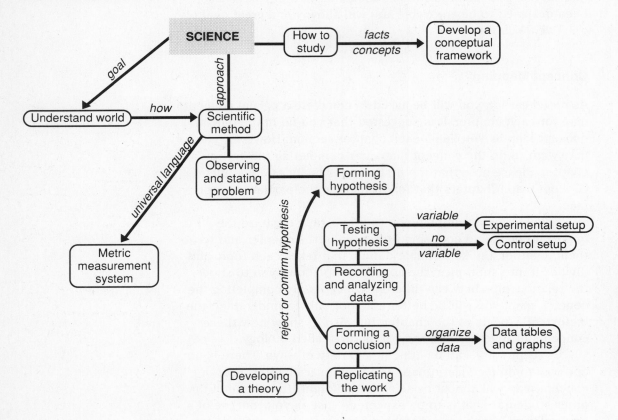

To emphasize the fact that concept maps can vary, the authors of this Study Guide have tried to vary the partial concept map provided for you in each chapter. However, there are some format rules you may want to follow. On your concept map, you will probably want to place the main concepts from the chapter (often only one, never more than three) in a box. The letters for these main concepts should be all capitalized. Leading away from these main concept boxes are lines, called leader lines, that connect the main concepts to subconcepts. Subconcepts are generally placed in ovals or circles rather than boxes. In addition, subconcepts are not all capitalized.

When a process, fact, or application can be used to tie together concepts in boxes or ovals, an arrow is usually drawn from one concept to another rather than just a straight leader line. Along the arrow is written the process, fact, or application that shows how the concepts are related.

## Concept Map 1A

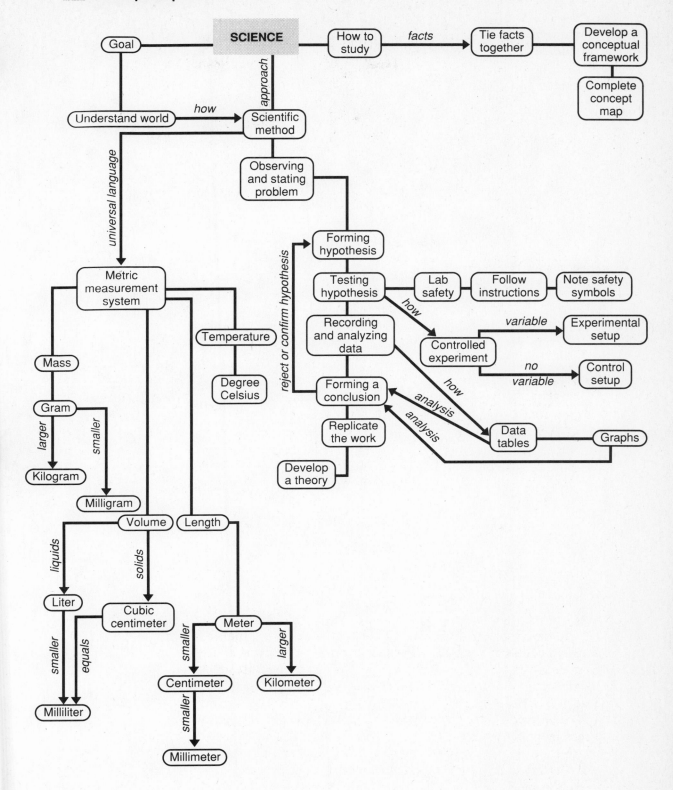

# S T U D Y
# G U I D E

| Section 1–1 | **What Is Science?** | *(pages 5–6)* |

## SECTION REVIEW

In this section you were introduced to the term *science*. You learned that there are many ways in which people can explain the world about them. Some explanations are classified as myths and legends. Myths and legends try to explain events in nature, but they are not scientific explanations.

Like myths and legends, science attempts to explain events that occur in our world. The process of science, however, does more than simply providing possible explanations of the events and observations that make up life. For an explanation to be considered scientific, a specific process must be followed. First of all, science assumes that there are natural causes for everything that occurs in nature. Second, science uses observations and experiments to test the possible causes for an event in nature. It is only through testing and experimentation that an explanation for an event can be considered a scientific explanation.

### Applying Definitions: Building Vocabulary Skills

**1.** Using your own words, define the term *science*. _____

_____

_____

_____

_____

**2.** What is the goal of science? _____

_____

_____

_____

_____

### Relating Concepts: Finding the Main Ideas

Listed below are a series of events, observations, statements, or questions. Determine whether each item qualifies as science or an area that science would study. Put an *S* next to each item that qualifies as an area of scientific study. If you do not think an item qualifies as science, write in the field of study in which that item best fits.

**1.** Why is the sky blue? _____

**2.** 2 + 2 = 4 _____

**3.** The sun rises in the east and sets in the west. _____

**4.** Monet was the most prominent of the French Impressionists. _____

**5.** On a clef scale, the note directly to the right of E is F. _____

**6.** "I think, therefore I am." _____

**7.** Life on Earth probably began in ancient oceans. _____

**8.** The closest planet to our sun is Mercury. _____

**9.** "No man is an island unto himself." _____

**10.** The sun is the primary source of energy for living things. _____

**11.** In a right triangle, the sum of the squares of the sides is equal to the square of the hypotenuse. _____

**12.** Organisms made up of more than one cell are called multicellular organisms.

_____

**13.** $a^2 + b^2 = c^2$ _____

**14.** Evolution can be defined as a change in living things over time. _____

**15.** Shakespeare is often considered the greatest writer in the English language.

_____

## Concept Mapping

The construction of and theory behind concept mapping are discussed on pages vii–ix in the front of this Study Guide. Read those pages carefully. Then consider the main concepts in Section 1–1 and how you would organize them in a concept map. Now look at the completed concept map for Chapter 1 on page 12. Notice how the concepts in Section 1–1 were incorporated into the concept map. In future chapters you will be asked to fill in your own concept map.

**Section 1–2** | **The Scientific Method** | *(pages 7–15)*

### SECTION REVIEW

The scientific method is the name given to the process that is used to both propose and test explanations about events that occur in nature. In this section you were introduced to the basic steps of the scientific method. You learned that the first step is to identify or observe a problem. Next, a hypothesis, or a suggested answer to the problem, is formulated. If the hypothesis appears reasonable, the next step in the scientific method is to test the hypothesis. Most hypotheses are tested through experimentation. During the experiment, results and observations are carefully recorded. The next step is to analyze the results. However, even if the results of an experiment seem to confirm the hypothesis, the hypothesis is not considered confirmed until the experiment is run many times and can be repeated, or replicated, by other scientists.

In order to confirm a hypothesis through an experiment, both a control setup and an experimental setup must be established. The experimental setup contains a variable that is being tested. The control setup is run exactly like the experimental setup but without the variable being tested. In this way scientists can be sure that the results of the experiment are due to the variable and not to some hidden factor that was not even considered.

Because scientists must be able to replicate the work of other scientists and communicate with each other, a universal system of measurement is used. This measurement system, which is based on powers of 10, is called the metric system. The basic unit of length in the metric system is the meter, and the basic unit of mass is the kilogram. The basic unit of volume for liquids is the liter. However, the cubic centimeter is used when giving the volume of solids in the metric system. Temperature in the metric system is measured on the Celsius scale.

### The Scientific Method: Using the Main Ideas

Describe an experiment you could run to test the following hypothesis. Keep in mind that any experiment must have both a control setup and an experimental setup, and that only one variable at a time should be tested.

*Hypothesis:* Turtle eggs develop into male turtles when exposed to cold temperatures and into female turtles when exposed to warm temperatures.

*Experiment:*

_____

_____

_____

_____

_____

_____

What was the variable in your experiment? _____

### Graphing Data

In the laboratory, yeast cells are usually grown at a temperature of 25°C. A scientist wanted to determine if yeast cells would reproduce faster or slower when the temperature was dropped to 15°C. The scientist performed an experiment and recorded the data in data tables. Examine the data tables for this experiment. Then plot the data for both the control setup and the experimental setup on the graph paper. In this graph the horizontal and vertical axes have been established for you. In future graphs, however, you will have to determine the organization of the axes on a graph.

**GROWTH OF YEAST CELLS**

**Temperature at 15°C (experimental setup)**

| Time (hours) | 0 | 2 | 4 | 6 | 8 | 10 | 12 | 14 | 16 | 18 |
|---|---|---|---|---|---|---|---|---|---|---|
| Number of yeast cells | 1 | 4 | 20 | 38 | 90 | 150 | 220 | 350 | 430 | 550 |

**Temperature at 25°C (control setup)**

| Time (hours) | 0 | 2 | 4 | 6 | 8 | 10 | 12 | 14 | 16 | 18 |
|---|---|---|---|---|---|---|---|---|---|---|
| Number of yeast cells | 1 | 26 | 74 | 170 | 330 | 550 | 600 | 650 | 665 | 670 |

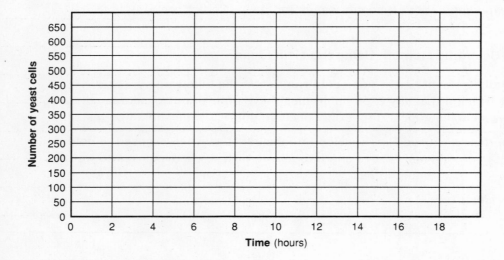

Based on the graphs, what conclusion can you draw about the growth and reproduction of

yeast cells? _____

_____

_____

_____

### ▨ Find the Oddball: Building Vocabulary Skills

Look over the following lists of metric units. In each list, one of the units does not belong. On the line below the list, write the unit that does not belong and explain your choice.

**1.** milliliter, gram, liter, cubic centimeter _____

_____

_____

_____

_____

**2.** meter, angstrom, kilometer, milliliter _____

_____

_____

_____

_____

**3.** micrometer, kilogram, metric ton, milligram _____

_____

_____

_____

_____

**4.** nanometer, angstrom, degree Celsius, centimeter _____

_____

_____

_____

_____

### ▨ Concept Mapping

The construction of and theory behind concept mapping are discussed on pages vii–ix in the front of this Study Guide. Read those pages carefully. Then consider the main concepts in Section 1–2 and how you would organize them in a concept map. Now look at the completed concept map for Chapter 1 on page 12. Notice how the concepts in Section 1–2 were incorporated into the concept map. In future chapters you will be asked to fill in your own concept map.

| Section 1-3 | Science "Facts" and "Truth" | (pages 15-18) |

## SECTION REVIEW

In order to study biology, or any science, it is important to learn all the key facts and bits of information provided by your textbook and your teacher. However, if you do not tie the information together, you will find that you have accumulated a great amount of data but you will not be able to relate one fact to another. There are basic underlying concepts in biology. Grouping the facts you learn into a conceptual framework will enable you to learn faster, remember more, and even get a better grade. And it will help you to place the new facts and information you gain in future years into your conceptual framework.

An understanding of both facts and basic concepts is not just important in school, it is a vital part of becoming a good citizen. As a voter, you will be asked to decide on many questions regarding scientific research, the environment, medical care, and so on. If you do not have a solid understanding of science, you will have to rely on the advice of scientists and those who understand science.

### Recognizing Fact and Opinion: Using the Main Ideas

Each of the following situations could occur. In each case, examine the situation and try to determine the reaction of each of the groups listed below the situation. Keep in mind that each group has "all the facts" but that interpretations and values might vary from one group to another.

*Situation 1:* The United States government decides to sell off several large areas now located in national parks. These areas can be mined for coal and various precious minerals. How might the following groups react to this proposal to sell national parklands in order to obtain important resources?

Coal miners' union: _____

_____

_____

_____

Park forest rangers: _____

_____

_____

_____

Camping associations: _____

_____

_____

_____

**Situation 2:** NASA has asked for funding to build a space station that will orbit the Earth. If money is spent on the space station, it will not be available for other purposes. How might the following groups react to NASA's request for funding?

Amateur astronomers: _____

_____

_____

Senate budget committee: _____

_____

_____

Aerospace industry: _____

_____

_____

**Situation 3:** A dam is about to be built in an area that badly needs electric power. Just before construction begins, however, a small worm is found living in an area that will be flooded by the dam. Scientists discover that this particular worm lives only in this one spot on Earth. If the dam is built, the worm will become extinct. If the dam is not built, progress in the area will be halted due to lack of electric power. How might the following groups react to this dilemma?

Small businesses in the area:_____

_____

_____

Conservation groups: _____

_____

_____

You: _____

_____

_____

### Concept Mapping

The construction of and theory behind concept mapping are discussed on pages vii–ix in the front of this Study Guide. Read those pages carefully. Then consider the main concepts in Section 1–3 and how you would organize them in a concept map. Now look at the completed concept map for Chapter 1 on page 12. Notice how the concepts in Section 1–3 were incorporated into the concept map. In future chapters you will be asked to fill in your own concept map.

| Section 1–4 | Safety In the Laboratory | *(pages 18–20)* |

## SECTION REVIEW

Biology, like any science, is not just to be read about in books. It is more exciting and valuable when biological facts and concepts are developed through laboratory work. Although laboratory work can be both fun and informative, there is always the danger that an accident may occur. All of the investigations you will do this year are safe when performed properly. However, even the safest experiment can prove hazardous when you do not follow the basic rules of safety. The most important thing to remember during your year in biology is to follow your teacher's directions or the directions in your book exactly. Do not improvise without permission. Chemicals that may seem safe, for example, could become highly toxic, or poisonous, when mixed improperly. So it is vital that you follow all directions exactly. And when you are not sure what to do, ask. Don't try something to "see what happens." If you follow these basic rules and learn the meaning of the laboratory safety symbols shown in Figure 1–16 in your textbook, you will not only have a productive year in the laboratory, you will have a safe year as well.

### Applying Safety Rules: Interpreting the Main Ideas

Study each of the following illustrations. In the space provided to the right of each drawing, write any safety rules you think are being broken and explain why the rule is being broken.

_____
_____
_____
_____
_____
_____
_____
_____

_____
_____
_____
_____
_____
_____
_____

_____
_____
_____
_____
_____
_____
_____

_____
_____
_____
_____
_____
_____

_____
_____
_____
_____
_____
_____

### Concept Mapping

The construction of and theory behind concept mapping are discussed on pages vii–ix in the front of this Study Guide. Read those pages carefully. Then consider the main concepts in Section 1–4 and how you would organize them in a concept map. Now look at the completed concept map for Chapter 1 on page 12. Notice how the concepts in Section 1–4 were incorporated into the concept map. In future chapters you will be asked to fill in your own concept map.

| Section 1-5 | **The Spaceship Called Earth** | *(pages 20–21)* |

## SECTION REVIEW

In the past, Earth was considered boundless, with unlimited resources and space. Today, we know that there is a limit to the vital resources on Earth, as well as a limit to the amounts of food, oxygen, and clean water that are available to living things. For this reason, many scientists have suggested treating our planet much like a giant spaceship traveling through space. However, unlike other spaceships, planet Earth cannot replenish supplies when its vital resources run low. So we must make every effort to use our resources wisely and to conserve valuable resources whenever possible.

### Calculating Water Use: Relating the Main Ideas

One of our most precious resources is water. Earth is the only known planet to have large amounts of liquid water. The diagram below shows the average amount of water used for various purposes every day by a family of four. Choosing from the list below, try to decide which of the items on the list should be placed on the appropriate Use line in the diagram. This will show you the number of liters of water used for each activity on the list. Next, convert the number of liters of water to a percentage. Put that answer in the blank to the right of the diagram. **Note:** The first line is already filled in for you.

Watering lawn          Bathing               Washing car
Garbage disposal       Drinking and cooking  Dishwasher
Flushing toilet        Washing machine

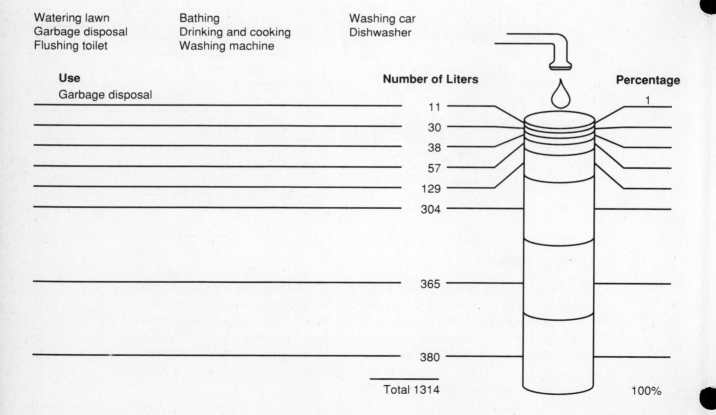

| Use | Number of Liters | Percentage |
|-----|------------------|------------|
| Garbage disposal | 11 | 1 |
| | 30 | |
| | 38 | |
| | 57 | |
| | 129 | |
| | 304 | |
| | 365 | |
| | 380 | |
| | **Total 1314** | **100%** |

## Using the Writing Process

Use your writing skills and imagination to respond to the following writing assignment. You will probably need an additional sheet of paper to complete your response.

The planet Kufu is very much like our planet. On Kufu, however, all activities are governed by the scientific method. Anything that cannot be tested by the scientific method is not tolerated on this planet.

You have been sent to write a news report on life on Kufu. This report should be a critical analysis of the lifestyle on this planet.

*Hint:* Before starting your news assignment, make a list of the different areas of society that affect everyday life—such as politics, science, economics, and interpersonal relationships. Note how each of these areas would be positively and/or negatively affected by the exclusive use of the scientific method.

_____

_____

_____

_____

_____

_____

_____

_____

_____

_____

_____

_____

_____

_____

_____

_____

_____

## Concept Mapping                                                    *Chapter 1*

The concept map below has been done for you for Chapter 1. Study it to see how it was completed. There are other ways, of course, to concept map Chapter 1. You may want to practice concept mapping by developing an alternate concept map for Chapter 1.

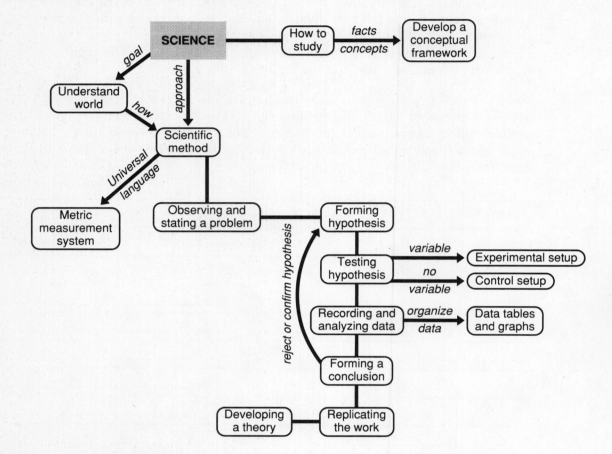

© Prentice-Hall, Inc.

# STUDY GUIDE

**Section 2–1     Characteristics of Living Things**                    *(pages 27–31)*

### SECTION REVIEW

In this section you learned that all living things share certain characteristics. Scientists generally agree that all living things (1) are made up of cells, (2) reproduce, (3) grow and develop, (4) obtain and use energy, and (5) respond to their environment.

As you considered each of these characteristics in detail, you learned that organisms can be made up of a single cell or of many cells. You also learned that reproduction can be sexual or asexual. You discovered that living things go through a predictable life cycle that includes growth, development, aging, and death. You learned that living things obtain and use energy in the processes of anabolism and catabolism. You also discovered that organisms respond to internal and external stimuli in order to improve their chances for survival. The ability of an organism to do this is called homeostasis.

### Understanding Relationships: Building Vocabulary Skills

Tell how the words in each pair or group are related.

**1.** unicellular, multicellular _____

_____

_____

_____

**2.** anabolism, catabolism, metabolism _____

_____

_____

_____

**3.** homeostasis, stimulus _____

_____

_____

_____

**4.** sexual reproduction, asexual reproduction _____

_____

_____

_____

### Relating Concepts: Finding the Main Ideas

Listed below are the five basic characteristics of living things. Each of the statements that follow describes either a living or a nonliving thing. If the statement describes a living thing, place in the blank before the statement the letter of the characteristic to which the statement refers. If the statement describes a nonliving thing, place an *N* in the blank.

**A.** Made up of cells  
**B.** Reproduce  
**C.** Grow and develop  
**D.** Obtain and use energy  
**E.** Respond to their environment

_____ **1.** A butterfly emerges from a cocoon.

_____ **2.** A lizard sleeps in the sun to obtain warmth.

_____ **3.** A runner eats a spaghetti dinner the night before a race.

_____ **4.** Rust forms on an iron nail.

_____ **5.** Wood is made up of the cell walls that separated tree cells.

_____ **6.** A piece of paper is cut in half to form two smaller pieces.

_____ **7.** A bird lays an egg that will hatch into a baby bird.

_____ **8.** A tadpole gradually changes into a frog.

_____ **9.** Birds fly south to find food in the winter.

_____ **10.** A green plant converts water and carbon dioxide into sugar.

_____ **11.** A spider lays its eggs in an egg sac.

_____ **12.** A bacterium divides to form two bacteria.

### Life Span of an Organism

The diagram below shows various stages in the life span of an organism. Use the diagram to answer the questions that follow.

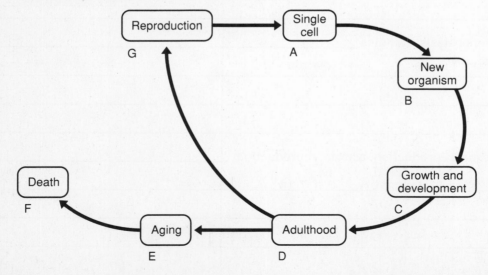

© Prentice-Hall, Inc.

1. During what period in its life span does an organism reproduce?

   _____

   _____

2. For humans and other mammals, which point on the diagram represents fertilization?

   Which point represents birth? _____

   _____

3. How would the transition from point A to point B vary for different types of animals?

   _____

   _____

   _____

   _____

4. Which periods in a life span might tend to overlap? Why? _____

   _____

   _____

5. Describe what happens to an organism as it moves from point D to point E on the

   diagram. _____

   _____

   _____

6. Is it possible for an organism to move to point F without passing through the other

   stages? Explain your answer. _____

   _____

   _____

   _____

## Concept Mapping

The construction of and theory behind concept mapping are discussed on pages vii–ix in the front of this Study Guide. Read those pages carefully. Then consider the concepts presented in Section 2–1 and how you would organize them into a concept map. Now look at the concept map for Chapter 2 on page 20. Notice that the concept map has been started for you. Add the key facts and concepts you feel are important for Section 2–1. When you have finished the chapter, you will have a completed concept map.

**Section 2–2**   **Biology: The Study of Life**    *(pages 31–39)*

## SECTION REVIEW

Biology means the study of life. A biologist is anyone who uses the scientific method to study living things. In this section you learned that the field of biology consists of many branches. Some of these branches are quite general, such as botany, which is the study of plants, or zoology, which is the study of animals. Other branches are more specialized, such as paleontology, which is the study of extinct organisms.

You learned that biologists use a wide variety of scientific tools. Many of these tools, such as scales and graduated cylinders, are commonly found in all science laboratories. One tool, however, is especially important to the study of biology. This tool is the microscope. Biologists use several different types of microscopes. The one you are probably most familiar with is the compound light microscope. An important group of microscopes used to study very small objects are the electron microscopes. Two types of electron microscopes are the transmission electron microscope (TEM) and the scanning electron microscope (SEM).

In the last part of this section you learned about the laboratory techniques of a biologist. Some of the more common techniques include staining, centrifugation, microdissection, and cell cultures.

**Understanding Prefixes and Suffixes: Building Vocabulary Skills**

The chart below shows the meanings of the prefixes and suffixes that make up biological terms. Following the chart is a list of definitions. Use the chart to form each term that is defined.

| Prefix | Meaning | Prefix | Meaning | Suffix | Meaning |
|--------|---------|--------|---------|--------|---------|
| anti– | against | herb– | pertaining to plants | –cyst | pouch |
| arth– | joint, jointed | hetero– | different | –derm | skin, layer |
| auto– | self | homeo– | same | –gen | producing |
| bio– | related to life | macro– | large | –itis | inflammation |
| chloro– | green | micro– | small | –logy | study |
| cyto– | cell | multi– | consisting of many units | –meter | measurement |
| di– | double | osteo– | bone | –osis | condition, disease |
| epi– | above | photo– | pertaining to light | –phase | stage |
| exo– | outer, external | plasm– | forming substance | –phage | eater |
| gastro– | stomach | proto– | first | –pod | foot |
| hemo– | blood | syn– | together | –stasis | stationary condition |

1. Top skin layer: _____

2. The same condition: _____

3. Instrument for measuring light: _____

4. Producing against: _____

**5.** Inflammation of the stomach: _____

**6.** Study of bones: _____

**7.** Instrument for measuring very small distances: _____

**8.** Cell disease: _____

**9.** Large eater: _____

**10.** Stomach-foot: _____

### Branches of Biology: Applying the Main Ideas

The diagram below shows some of the branches, or divisions, of biology organized in a biological tree. Each of the questions that follow would be studied by a biologist in a particular branch. Read each question and decide which type of biologist would be most likely to study it. Then write the number of the question in the blank on the correct branch.

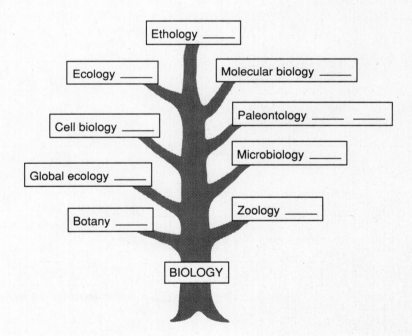

**1.** How do malignant tumor cells differ from benign tumor cells?

**2.** How does a bear prepare for hibernation?

**3.** How did dinosaurs change over time?

**4.** What special behavior does an albatross exhibit before mating?

**5.** How do drugs cause changes in DNA?

**6.** How is plant growth affected by changes in light and temperature?

**7.** What were the first horses like?

**8.** How is loss of forest land affecting the Earth's climate?

**9.** What is the effect of acid rain on organisms living in Blue Lake?

**10.** How does a virus reproduce?

## Compound Light Microscope

The diagram below shows a typical compound light microscope. The list that follows names each of the labeled parts. Identify the parts by placing the correct letter in the blank before each part.

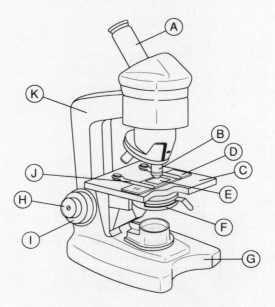

_____ Stage clips          _____ Glass slide

_____ Fine adjustment knob          _____ Stage

_____ Ocular lens (eyepiece)          _____ Coverslip

_____ Base          _____ Coarse adjustment knob

_____ Objective lens          _____ Arm

_____ Diaphragm

## Concept Mapping

The construction of and theory behind concept mapping are discussed on pages vii–ix in the front of this Study Guide. Read those pages carefully. Then consider the concepts presented in Section 2–2 and how you would organize them into a concept map. Now look at the concept map for Chapter 2 on page 20. Notice that the concept map has been started for you. Add the key facts and concepts you feel are important for Section 2–2. When you have finished the chapter, you will have a completed concept map.

## Using the Writing Process

Use your writing skills and imagination to respond to the following writing assignment. You will probably need an additional sheet of paper to complete your response.

A miniature alien has been sent to Earth to bring a sample of life back to her home planet. After looking around, the alien finds some apple seeds. She decides that they would be much easier to carry back than most of the living things she has seen. However, the alien is not sure if the apple seeds are alive. How could she find out if they are alive? What suggestions would you give to the alien?

_____

_____

_____

_____

_____

_____

_____

_____

_____

_____

_____

_____

_____

_____

_____

_____

_____

_____

_____

**Concept Mapping**                                                    *Chapter 2*

The concept map below has been started for you. Use this partial concept map to complete the concept map for each section of the chapter. When you are done, you will have a concept map for the entire chapter.

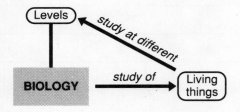

STUDY
GUIDE

**Section 3–1**  **Nature of Matter**  *(pages 45–47)*

### SECTION REVIEW

In this section you learned that matter is anything that has mass and volume. Mass is the quantity of matter in an object. Volume is the amount of space matter takes up.

Matter is described and identified by characteristics you can observe. These characteristics are divided into two general categories: physical properties and chemical properties.

Physical properties can be observed and measured without changing the identity of matter. Mass, volume, color, odor, shape, texture, taste, hardness, melting point, boiling point, and phase are some examples of physical properties.

Chemical properties describe the ability of matter to change from one substance into another as the result of a chemical change. One example of a chemical property is the ability to burn.

### Relating Terms: Building Vocabulary Skills

Using your own words, define each member of the following pairs of terms.
Then explain how the paired terms are related to one another.

**1.** matter, volume _____

_____

**2.** mass, weight _____

_____

_____

**3.** physical property, chemical property _____

_____

_____

_____

**4.** chemical change, phase change _____

_____

_____

_____

**5.** solid, gas _____

_____

### Relating Concepts: Using the Main Ideas

Each of the following illustrations shows a change in matter. In the space provided, identify each change as physical or chemical. Explain your response.

**1.** Wood burns: _____

_____

_____

_____

_____

_____

_____

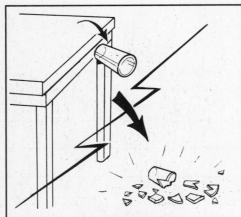

**2.** Glass breaks: _____

_____

_____

_____

_____

_____

_____

**3.** Ice melts: _____

_____

_____

_____

_____

_____

_____

**4.** Bicycle rusts: _____

_____

_____

_____

_____

_____

_____

_____

_____

**5.** Your just-washed hair dries: _____

_____

_____

_____

_____

_____

_____

_____

_____

### Concept Mapping

The construction of and theory behind concept mapping are discussed on pages vii–ix in the front of this Study Guide. Read those pages carefully. Then consider the concepts presented in Section 3–1 and how you would organize them into a concept map. Now look at the concept map for Chapter 3 on page 32. Notice that the concept map has been started for you. Add the key facts and concepts you feel are important for Section 3–1. When you have finished the chapter, you will have a completed concept map.

**Section 3–2**  **Composition and Matter**  *(pages 47–51)*

## SECTION REVIEW

In this section you learned that the basic unit of matter is the atom. The atom is made up of even smaller particles. Three of these subatomic particles are the proton, the neutron, and the electron. Both the positively charged protons and the electrically neutral neutrons form the nucleus of the atom. The negatively charged electrons travel around the nucleus of the atom in a series of distinct energy levels.

The number of protons in the nucleus of an atom, or the atomic number, is used to identify the atom. The total number of protons and neutrons in the nucleus is called the mass number.

Substances that consist entirely of only one type of atom are called elements. Each element is represented by a chemical symbol. The number of protons in an atom of an element never changes. The number of neutrons, however, may vary from one atom of an element to the next. Atoms of the same element that have different numbers of neutrons are called isotopes. Certain kinds of isotopes are radioactive.

Substances that consist of more than one type of atom are called chemical compounds. Most of the materials in the living world are compounds.

### Match the Matter: Building Vocabulary Skills

Match the numbered terms with their definitions. Write the letter of the definition that corresponds with each term in the space provided.

_____ 1. mass number

**a.** atoms of the same element with different numbers of neutrons

_____ 2. chemical symbol

**b.** positively charged particle

_____ 3. isotopes

**c.** subatomic particle located outside the nucleus of the atom

_____ 4. atomic number

**d.** substance that contains two or more different kinds of atoms.

_____ 5. nucleus

**e.** number of protons in an atom

_____ 6. neutron

**f.** shorthand way of representing an element

_____ 7. element

**g.** center of the atom

_____ 8. proton

**h.** electrically neutral subatomic particle

_____ 9. chemical compound

**i.** number of protons plus the number of neutrons in an atom

_____ 10. electron

**j.** substance consisting of only one type of atom

### Making Calculations: Relating the Main Ideas

Atoms that have the same number of protons and electrons are electrically neutral. However, atoms may gain or lose electrons during chemical reactions. This creates an imbalance of negative and positive charges. Atoms may have a negative charge because they have gained extra electrons. Such atoms are called negative ions. Other atoms may have a positive charge because they have lost electrons. These atoms are called positive ions.

The following table contains information about several atoms. Using what you have learned in the preceding paragraph and in Section 3–2, complete the table. Enough information has been provided for you to fill in all the blanks.

| Element | Atomic Number | Mass Number | Number of Protons | Number of Neutrons | Number of Electrons | Isotope, Ion, or Neutral Atom |
|---|---|---|---|---|---|---|
| Aluminum (Al) | 13 | 27 | | | 13 | neutral atom |
| Bromine (Br) | | | 35 | 45 | 36 | |
| Carbon (C) | 6 | | | 6 | 6 | |
| Carbon (C) | 6 | 14 | | | 6 | |
| Helium (He) | 2 | 4 | | | | neutral atom |
| Hydrogen (H) | 1 | | 1 | | | neutral atom |
| Hydrogen (H) | | 1 | | | 0 | |
| Lithium (Li) | 3 | 7 | | | 2 | |
| Nitrogen (N) | | 14 | | | | neutral atom |
| Oxygen (O) | | 18 | 8 | | 8 | |
| Oxygen (O) | 8 | 16 | | | 6 | |
| Potassium (K) | | 39 | 19 | | | neutral atom |

### Concept Mapping

The construction of and theory behind concept mapping are discussed on pages vii–ix in the front of this Study Guide. Read those pages carefully. Then consider the concepts presented in Section 3–2 and how you would organize them into a concept map. Now look at the concept map for Chapter 3 on page 32. Notice that the concept map has been started for you. Add the key facts and concepts you feel are important for Section 3–2. When you have finished the chapter, you will have a completed concept map.

| Section 3–3 | Interactions of Matter | *(pages 52–54)* |

## SECTION REVIEW

Chemical compounds are formed when atoms of different elements combine in a process known as chemical bonding. As you learned in this section, the formation of a chemical bond depends upon the number of electrons in the outermost energy level around the nucleus of an atom. Each energy level can hold only a certain number of electrons. Atoms are stable when their outermost energy level is filled. An atom will gain, lose, or share electrons in order to become stable.

Atoms change their physical and chemical properties when they form a compound. An ionic bond results when electrons are transferred from one atom to another. The atoms that gain electrons when an ionic bond is formed are called negative ions. The atoms that lose electrons are called positive ions.

If atoms share electrons in order to achieve stability, a covalent bond is formed. The combination of atoms that results from a covalent bond is a unit called a molecule.

### Applying Definitions: Building Vocabulary Skills

Use the following terms to label the structures shown in each of the diagrams: electron, nucleus, neutral atom, positive ion, negative ion, ionic bond, covalent bond, molecule. **Note:** Some of the labels may be used more than once on a diagram. Some labels may not be used at all.

After you have labeled a diagram, answer the questions that follow it.

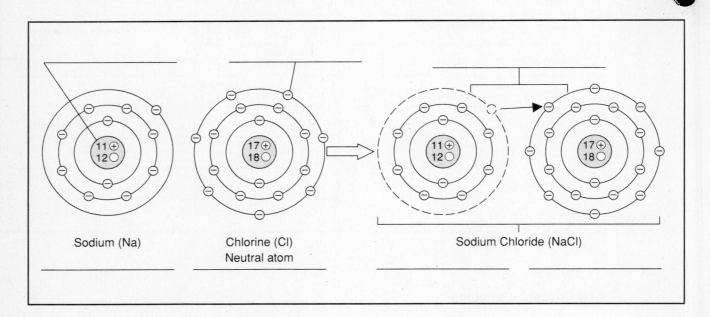

Sodium (Na)

Chlorine (Cl)
Neutral atom

Sodium Chloride (NaCl)

**1.** What basic process is shown in the diagram? Explain. _____

_____

**2.** How many electrons can the innermost energy level hold? _____

The outermost? _____

**3.** What kind of bond is shown? Explain. _____

_____

_____

**4.** How does an ion differ from a neutral atom? _____

_____

_____

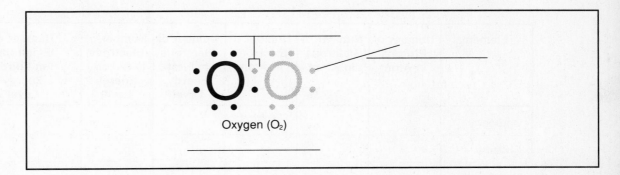

Oxygen (O₂)

**5.** What do the dots in the above diagram represent? _____

_____

**6.** What kind of bond is shown? Explain. _____

_____

_____

### Analyzing Data: Using the Main Ideas

The accompanying diagram shows a small portion of the periodic table of
the elements. Use the information in the diagram to answer questions 1
through 3 and to fill in the table that follows.

| | | | | | | | |
|---|---|---|---|---|---|---|---|
| 1<br>**H**<br>Hydrogen<br>1 | | | | | | | 2<br>**He**<br>Helium<br>4 |
| 3<br>**Li**<br>Lithium<br>7 | 4<br>**Be**<br>Beryllium<br>9 | 5<br>**B**<br>Boron<br>11 | 6<br>**C**<br>Carbon<br>12 | 7<br>**N**<br>Nitrogen<br>14 | 8<br>**O**<br>Oxygen<br>16 | 9<br>**F**<br>Fluorine<br>19 | 10<br>**Ne**<br>Neon<br>20 |
| 11<br>**Na**<br>Sodium<br>23 | 12<br>**Mg**<br>Magnesium<br>24 | 13<br>**Al**<br>Aluminum<br>27 | 14<br>**Si**<br>Silicon<br>28 | 15<br>**P**<br>Phosphorus<br>31 | 16<br>**S**<br>Sulfur<br>32 | 17<br>**Cl**<br>Chlorine<br>35 | 18<br>**Ar**<br>Argon<br>40 |

1. What is the boldfaced letter or letters in the center of the box?

   _____

2. What is the number above a boldfaced letter or letters? _____

   _____

3. What is the number below the name of an element? _____

   _____

| Element | Number of Protons in Atom | Number of Neutrons in Atom | Number of Electrons in Atom | Number of Electrons in First Energy Level | Number of Electrons in Second Energy Level | Number of Electrons in Third Energy Level |
|---|---|---|---|---|---|---|
| Carbon | | | | | | |
| Chlorine | | | | | | |
| Helium | | | | | | |
| Hydrogen | | | | | | |
| Lithium | | | | | | |
| Magnesium | | | | | | |
| Neon | | | | | | |
| Nitrogen | | | | | | |
| Oxygen | | | | | | |
| Phosphorus | | | | | | |
| Sodium | | | | | | |
| Sulfur | | | | | | |

## Concept Mapping

The construction of and theory behind concept mapping are discussed on pages vii–ix in the front of this Study Guide. Read those pages carefully. Then consider the concepts presented in Section 3–3 and how you would organize them into a concept map. Now look at the concept map for Chapter 3 on page 32. Notice that the concept map has been started for you. Add the key facts and concepts you feel are important for Section 3–3. When you have finished the chapter, you will have a completed concept map.

**Section 3–4** | **Chemical Reactions** | *(pages 55–56)*

## SECTION REVIEW

Any process in which a chemical change occurs is known as a chemical reaction. As you learned in this section, a chemical reaction converts elements or compounds known as reactants into elements or compounds known as products. A chemical equation uses symbols and formulas to describe a chemical reaction.

Energy is the most important factor in determining whether a reaction will occur. A chemical reaction that releases energy may occur spontaneously. A chemical reaction that requires energy, however, will not flow without a source of energy.

### Describing Chemical Reactions: Using the Main Ideas

A number of chemical reactions are described below. In the spaces provided, identify the reactant(s) and the product(s). Then write an equation for each reaction.

1. A sodium ion ($Na^+$) reacts with a chlorine ion ($Cl^-$) to form the compound sodium chloride (NaCl)

   Reactant(s): _____

   Product(s): _____

   Equation: _____

2. Magnesium oxide (MgO) is produced when a magnesium ion ($Mg^{2+}$) reacts with an oxygen ion ($O^{2-}$)

   Reactant(s): _____

   Products(s): _____

   Equation: _____

Most chemical reactions and their equations are more complex than those in the two exercises you have just completed. The next few problems are a bit more tricky because you will be asked to balance the equations. In a balanced equation, the number of atoms of each element is the same on both sides of the equation.

Here is how you balance a chemical equation.

- Write a chemical equation with correct symbols and formulas.
- Count the number of atoms of each element on each side of the arrow.
- Balance atoms by using coefficients. Coefficients are numbers that indicate how many atoms or molecules of each substance are involved in the reaction. **Note:** *Only change coefficients when balancing an equation. Never change symbols or formulas.*
- Check your work by counting the atoms of each element.

**3.** Nitrogen ($N_2$) and hydrogen ($H_2$) form ammonia ($NH_3$).

Reactant(s): _____

Product(s): _____

Equation: _____

**4.** Carbonic acid ($H_2CO_3$) breaks down to form water ($H_2O$) and carbon dioxide ($CO_2$).

Reactant(s): _____

Product(s): _____

Equation: _____

**5.** When it is burned, natural gas ($CH_4$) combines with oxygen gas ($O_2$) to produce carbon dioxide ($CO_2$) and water ($H_2O$).

Reactant(s): _____

Product(s): _____

Equation: _____

## Concept Mapping

The construction of and theory behind concept mapping are discussed on pages vii–ix in the front of this Study Guide. Read those pages carefully. Then consider the concepts presented in Section 3–4 and how you would organize them into a concept map. Now look at the concept map for Chapter 3 on page 32. Notice that the concept map has been started for you. Add the key facts and concepts you feel are important for Section 3–4. When you have finished the chapter, you will have a completed concept map.

## Using the Writing Process

Use your writing skills and imagination to respond to the following writing assignment. You will probably need an additional sheet of paper to complete your response.

> The radioactive elements have been accused of creating enormous problems in society. Scientists have discovered a way to eliminate all radioactive elements from Earth. Before taking this final irreversible step, however, the radioactive elements will be given one last chance to defend themselves. Write a short play entitled "Radioactive Elements on Trial."

_____

_____

_____

_____

_____

_____

_____

_____

_____

_____

_____

_____

_____

_____

_____

_____

_____

_____

_____

_____

**Concept Mapping**                                                    *Chapter 3*

The concept map below has been started for you. Add the key facts and concepts for each section of the chapter to this partial concept map. When you are done, you will have a concept map for the entire chapter.

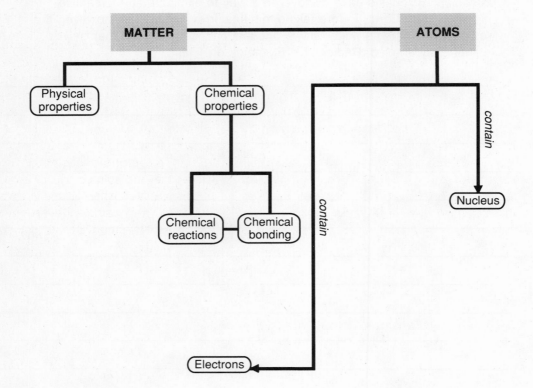

S T U D Y
G U I D E

| Section 4–1 | **Water** | *(pages 63–67)* |

## SECTION REVIEW

In this section you learned about the unique properties of water. The polarity of water molecules makes water especially good at forming mixtures. A mixture is a substance composed of two or more elements or compounds that are mixed but not chemically combined. There are two types of mixtures: solutions and suspensions.

Solutions are mixtures that consist of a solute and a solvent. A solute is the substance being dissolved. A solvent is the substance that does the dissolving. Water is the greatest solvent in the world. In a solution, the solute particles are uniformly spread throughout the solvent.

When ionically bonded compounds dissolve in water, they often dissociate into individual ions. Compounds that release hydrogen ions into solution are known as acids. Compounds that release hydroxide ions into solution are called bases. The pH scale indicates the relative concentrations of these two ions.

Suspensions are similar to solutions except that the solutes do not break into individual molecules when placed in water. Instead, they form small particles that remain suspended by the movement of the molecules of the solvent.

### ▓ Chemistry in Action: Building Vocabulary Skills

1. Substance A and substance B are poured into a beaker. They mix together but do not chemically combine. The result is called a (an)

   _____.

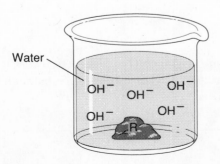

2. When substance R is placed in water, hydroxide ions are released into solution. Substance R must be a (an)

   _____.

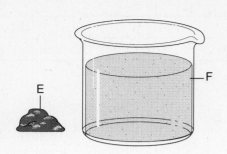

**3.** When substance E is placed in substance F, substance E breaks into small pieces that neither dissolve nor settle to the bottom of the container. This mixture is called a (an)

_____ .

**4.** Substance X has a pH value of 12. What does

that tell you about X? _____

_____

Substance Y has a pH value of 1. What do you

know about Y? _____

_____

If X and Y are mixed, they undergo a neutralization reaction. Explain what happens.

_____

_____

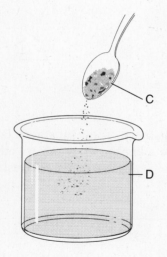

**5.** When substance C is placed in substance D, it dissolves and becomes uniformly spread throughout substance D. This mixture is

called a (an) _____ .

Substance C is the _____ .

Substance D is the _____ .

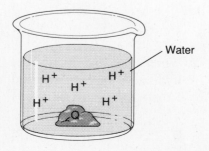

**6.** When substance Q is placed in water, it releases hydrogen ions into solution. Substance Q must be a (an)

_____ .

### pH and You: Relating Cause and Effect

Consider the following situation and answer the questions related to it. You wish to buy some fish for your new fish tank. The salesperson at the store tells you that different fish require different pH values for the water in which they live. Assuming your tap water has a pH value of 7.0, you purchase five Type A fish that require water with that pH value. After the first week, however, all of the fish die.

1. If you are sure that the water itself is the problem, and you know that there are no chemicals in the water that could have killed your fish, what are you to believe about

   your original assumption? _____

   _____

   You then buy two different types of fish and place five of each type in the tank. Type B fish require a pH range of 3.8 to 6.8. Type C fish require a pH range of 7.1 to 9.0. After another week, all Type C fish die, but all Type B fish live.

2. What does this tell you about your tap water? _____

   _____

   Knowing this, you purchase 15 more fish with the following pH requirements: Type D fish, 5.5 to 8.2; Type E fish, 6.0 to 9.0; Type F fish, 4.5 to 7.5. These fish continue to live for months. Then one day you notice that a container of liquid has accidentally spilled into the tank. You do not know what the liquid is or how it will affect your fish. Unfortunately, Type A fish and Type F fish soon die.

3. What must have happened to the water in the fish tank? _____

   _____

   _____

4. What does this tell you about the liquid that spilled into the tank?

   _____

   _____

5. What might this liquid have been? _____

### Concept Mapping

The construction of and theory behind concept mapping are discussed on pages vii–ix in the front of this Study Guide. Read those pages carefully. Then consider the concepts presented in Section 4–1 and how you would organize them into a concept map. Now look at the concept map for Chapter 4 on page 42. Notice that the concept map has been started for you. Add the key facts and concepts you feel are important for Section 4–1. When you have finished the chapter, you will have a completed concept map.

## SECTION REVIEW

The majority of naturally occurring compounds are inorganic compounds. Inorganic compounds are primarily those that do not contain carbon.

Organic compounds, on the other hand, contain carbon. Carbon has 4 electrons in its outermost energy level. Carbon is a unique element because of its ability to form covalent bonds that are strong and stable. In addition, carbon atoms can bond to other carbon atoms to form chains. The bonds between carbon atoms in these chains can be single, double, or triple covalent bonds, or combinations of these bonds. These chains can close to form ring structures containing single or double bonds, or a mixture of both. This gives even more variety to the kinds of molecules that carbon can form.

Many carbon-based compounds are formed by a chemical process known as polymerization. During this process, small compounds called monomers are joined together by chemical bonds to form large compounds called polymers. Very large polymers are called macromolecules. Polymerization results in great chemical diversity.

**Classifying Chemical Compounds**

Describe organic compounds. _____

_____

_____

Describe inorganic compounds. _____

_____

Using your definitions, identify the following compounds as either organic (O) or inorganic (I).

_____ $C_3H_8$          _____ $HBrO_3$          _____ $NH_3$          _____ $C_{10}H_{22}$

_____ $H_3PO_4$         _____ $HCl$             _____ $CH_4$          _____ $C_5H_{12}$

_____ $HClO_3$          _____ $CO_2$            _____ $NaCl$          _____ $H_2O$

**Constructing with Carbon: Using the Main Ideas**

Chemists use structural formulas to represent organic compounds. A structural formula indicates how many atoms of each element are present in a molecule, as well as how they are arranged. A structural formula uses chemical symbols for the elements and dashes to represent covalent bonds. For example, the structural formula for $CH_4$, or methane, is

1. Consider the following elements. Recalling what you learned about electrons and energy levels, indicate how many covalent bonds each element would form.

   Nitrogen (N)—atomic number 7          _____ bond(s)

   Hydrogen (H)—atomic number 1          _____ bond(s)

   Oxygen (O)—atomic number 8            _____ bond(s)

   Carbon (C)—atomic number 6            _____ bond(s)

2. Study the structural formulas that follow. Can these compounds exist as they appear here? Explain your answer.

   **a.** 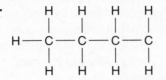     **b.** 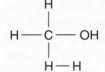     **c.**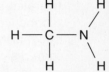

   a. _____

   b. _____

   c. _____

3. Now try your hand at drawing structural formulas. For each example, you are given enough information about the compound to draw the structural formula for a molecule.

   **a.** a straight-chain molecule containing 16 carbon atoms, 29 hydrogen atoms, and 1 carboxyl group, which has the structure $C \diagup^{O}_{OH}$

   **b.** $C_8H_{16}Cl_2$

## Concept Mapping

The construction of and theory behind concept mapping are discussed on pages vii–ix in the front of this Study Guide. Read those pages carefully. Then consider the concepts presented in Section 4–2 and how you would organize them into a concept map. Now look at the concept map for Chapter 4 on page 42. Notice that the concept map has been started for you. Add the key facts and concepts you feel are important for Section 4–2. When you have finished the chapter, you will have a completed concept map.

| Section 4–3 | Compounds of Life | (pages 70–77) |

## SECTION REVIEW

In this section you learned that organic compounds found in living things are classified into four groups: carbohydrates, lipids, proteins, and nucleic acids.

Carbohydrates are molecules that contain carbon, hydrogen, and oxygen atoms in an approximate ratio of 1:2:1 (C:H:O). Carbohydrates are divided into three major groups: monosaccharides, disaccharides, and polysaccharides. Monosaccharides combine to form disaccharides and polysaccharides in a process known as dehydration synthesis. The reverse process, in which polysaccharides are split apart to form monosaccharides, is called hydrolysis. Carbohydrates are the body's main source of energy.

Lipids also contain carbon, hydrogen, and oxygen, but in a different ratio than carbohydrates. Lipids store energy, form biological membranes, and serve as chemical messengers. Lipids are commonly known as fats, oils, and waxes. Generally, fats and waxes are solid at room temperature, whereas oils are liquids.

Proteins are polymers of amino acids, which are joined by peptide bonds. Proteins pump small molecules in and out of cells, are responsible for the ability of cells to move, and also help carry out chemical reactions through the actions of enzymes.

Nucleic acids are large, complex organic molecules that are polymers of nucleotides. There are two basic kinds of nucleic acids: ribonucleic acid (RNA) and deoxyribonucleic acid (DNA). Nucleic acids store and transmit the genetic information that is responsible for life.

### Science Writing: Building Vocabulary Skills

Below are several groups of related words. For each group, write a sentence or short paragraph that uses all the words. Pretend that you are explaining these words to someone who is totally unfamiliar with them.

**1.** Carbohydrates, polysaccharide, dehydration synthesis, hydrolysis, monosaccharide:

_____

_____

_____

_____

_____

**2.** Lipids, cholesterol, saturated, unsaturated, fatty acids, phospholipids, glycerol, sterols:

_____

_____

_____

_____

_____

**3.** Proteins, peptide bond, amino acid, polypeptide: _____

_____

_____

_____

_____

**4.** Enzyme, substrate, catalyst, active site: _____

_____

_____

_____

_____

**5.** Nucleic acids, ribonucleic acid, nucleotides, deoxyribonucleic acid:

_____

_____

_____

_____

### Be a Chemistry Detective

In each of the following exercises, you are given two clues about an organic compound. Using the knowledge you have gained in this section and your deductive abilities, determine the identity of the compound. Indicate your solution in the space provided.

**1.** • A polysaccharide.

• Animals store their excess sugar in this form.

_____

**2.** • Important in regulating chemical pathways, synthesizing materials needed by cells, and releasing energy.

• Speed up chemical reactions by binding to the reactants.

_____

**3.** • Complex molecules composed of carbon, oxygen, hydrogen, nitrogen, and phosphorous atoms.

• Polymers of molecules built up from a special 5-carbon sugar, a phosphate group, and a nitrogenous base.

_____

4. • A polysaccharide.

 • Gives strength and rigidity to plants.

 _____

5. • Consist of parts that dissolve well in water and parts that do not dissolve well in water.

 • Form liposomes when mixed with water.

 _____

6. • All have similar chemical structure but differ in a region known as the R group.

 • Can be joined by polymerization to form proteins.

 _____

7. • A polysaccharide.

 • Plants store excess sugar in this form.

 _____

8. • Contain carbon, hydrogen, and oxygen atoms in an approximate ratio of 1:2:1.

 • Has the formula $C_6H_{12}O_6$.

 _____

9. • Long chains of hydrogen and carbon atoms that have a carboxyl group attached at one end.

 • Combine with glycerol to form lipids.

 _____

10. • A disaccharide.

 • Other examples are maltose and lactose.

 _____

## Concept Mapping

The construction of and theory behind concept mapping are discussed on pages vii–ix in the front of this Study Guide. Read those pages carefully. Then consider the concepts presented in Section 4–4 and how you would organize them into a concept map. Now look at the concept map for Chapter 4 on page 42. Notice that the concept map has been started for you. Add the key facts and concepts you feel are important for Section 4–4. When you have finished the chapter, you will have a completed concept map.

## Using the Writing Process

Use your writing skills and imagination to respond to the following writing assignment. You will probably need an additional sheet of paper to complete your response.

> You are a scientist who, after laboring many years at your research, has just synthesized two new chemical compounds that when combined may cure the common cold. However, even under the most favorable conditions, the reaction proceeds too slowly to be effective. You are now looking for an enzyme to speed up the reaction. Write a job description for the enzyme you are hoping to find.

*Hint:* You may want to describe the two new chemical compounds first.

_____

_____

_____

_____

_____

_____

_____

_____

_____

_____

_____

_____

_____

_____

_____

_____

_____

_____

_____

**Concept Mapping**                                              *Chapter 4*

The concept map below has been started for you. Add the key factors and concepts for each section of the chapter to this partial map. When you are done, you will have a concept map for the entire chapter.

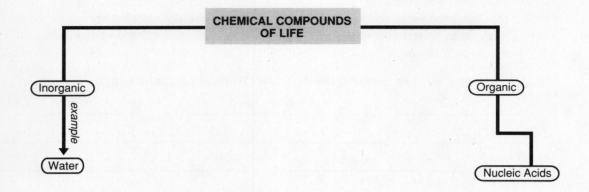

STUDY
GUIDE

| Section 5–1 | **The Cell Theory** | *(pages 87–88)* |

## SECTION REVIEW

In this section you read about some of the earliest efforts to examine the world of the very small. One of the most important events associated with these efforts was van Leeuwenhoek's invention of the microscope in the early 1600s. The microscope enabled people to see things they had never seen before. Among these things were the basic structures of living things: cells.

Further examination of cells by scientists such as Hooke, Brown, Schleiden, Schwann, and Virchow led to the formation of the cell theory. The cell theory forms the basic framework in which biologists have tried to understand living things. The cell theory states that living things are made of cells, that cells are the basic units of structure and function in living things, and that all cells come from preexisting cells.

### Defining Terms: Building Vocabulary Skills

In your own words, define each of the following terms. Use complete sentences.

**1.** Microscope: _____

_____

_____

_____

_____

**2.** Cells: _____

_____

_____

_____

**3.** Cell theory: _____

_____

_____

_____

_____

### Developing the Theory: Finding the Main Ideas

Explain how each of the following scientists contributed to the formation of the cell theory.

1. Anton van Leeuwenhoek: _____

_____

2. Robert Hooke: _____

_____

3. Robert Brown: _____

_____

4. Matthias Schleiden: _____

_____

5. Theodor Schwann: _____

_____

6. Rudolf Virchow: _____

_____

7. Now put it all together. Explain the cell theory in your own words. Note how the

contribution of each scientist relates to the theory. _____

_____

_____

_____

_____

_____

_____

_____

### Concept Mapping

The construction of and theory behind concept mapping are discussed on pages vii–ix in the front of this Study Guide. Read those pages carefully. Then consider the concepts presented in Section 5–1 and how you would organize them into a concept map. Now look at the concept map for Chapter 5 on page 54. Notice that the concept map has been started for you. Add the key facts and concepts you feel are important for Section 5–1. When you have finished the chapter, you will have a completed concept map.

**Section 5-2**   **Cell Structure**                                      *(pages 89–93)*

## SECTION REVIEW

Cells vary greatly in size and shape. But as you learned in this section, there are several structures that are common to most cells. The cells of plants, animals, and related organisms have three basic structures: the cell membrane, the nucleus, and the cytoplasm.

All cells are surrounded by a cell membrane. The cell membrane supports and protects the cell. It also regulates the passage of materials into and out of the cell. In some cells, such as those of plants and bacteria, the cell membrane is surrounded by a protective cell wall.

The nucleus is a large dark structure that serves as the control center of a cell. The nucleus consists of two membranes, called the nuclear envelope, that surround the cell's genetic information. This genetic information is contained in structures called chromosomes. In addition to the chromosomes, most nuclei contain a region called the nucleolus. Ribosomes, which are involved with the production of proteins, are made in the nucleolus. Remember that not all cells have a nucleus. Cells that contain a nucleus are said to be eukaryotic. Cells that do not are said to be prokaryotic.

The area between the nucleus and the cell membrane is called the cytoplasm. The cytoplasm contains many special structures that are involved in carrying out the cell's essential life functions.

**Identifying Cell Structures: Building Vocabulary Skills**

Label the parts of these typical cells. In the spaces provided beneath each cell, identify the cells as prokaryotic or eukaryotic.

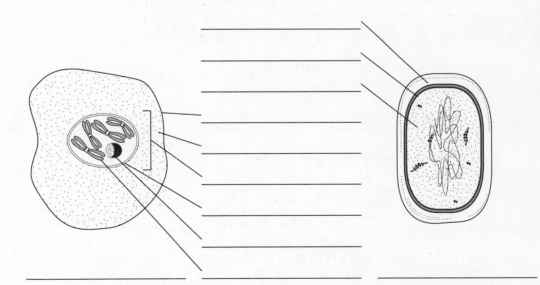

### Applying Concepts: Using the Main Ideas

Answer the following questions in your own words.

1. Name and give the functions of the three basic parts of a typical cell.

_____

_____

_____

_____

_____

2. What is the function of the proteins in the cell membrane?

_____

_____

_____

3. What is the function of the carbohydrates associated with the cell membrane?

_____

4. What is a cell wall? Name two kinds of cells that have cell walls.

_____

_____

_____

5. How are the nuclear envelope, nucleolus, and chromosomes related to one another?

Briefly describe each of these structures. _____

_____

_____

_____

_____

### Concept Mapping

The construction of and theory behind concept mapping are discussed on pages vii–ix in the front of this Study Guide. Read those pages carefully. Then consider the concepts presented in Section 5–2 and how you would organize them into a concept map. Now look at the concept map for Chapter 5 on page 54. Notice that the concept map has been started for you. Add the key facts and concepts you feel are important for Section 5–2. When you have finished the chapter, you will have a completed concept map..

**Section 5–3**  **Cytoplasmic Organelles**                                    *(pages 94–99)*

## SECTION REVIEW

Organelles are tiny structures found inside the cytoplasm of the cell. Each type of organelle performs a specialized function that helps maintain a cell's life. In this section you learned about many important organelles.

Mitochondria enable organisms to use the chemical energy stored in food. Chloroplasts trap the energy of sunlight and convert it into chemical energy. Ribosomes are the sites at which the cell makes proteins. The endoplasmic reticulum transports materials throughout the inside of the cell. The Golgi apparatus modifies, collects, packages, and distributes cell products. Lysosomes contain enzymes that are used to digest food particles, foreign materials, and even cells or organelles that are damaged or have outlived their usefulness. Vacuoles and plastids store materials. And the cytoskeleton supports cell structure and drives cell movement.

### Recognizing Function: Building Vocabulary Skills

From the following word bank, select the term that best fits each description. In the space provided, write the term you have selected. *Note:* Some terms may be used more than once. Some may not be used at all.

| Word Bank | | |
|---|---|---|
| Centriole | Golgi apparatus | Plastid |
| Chloroplast | Lysosome | Ribosome |
| Cilium | Microfilament | Rough ER |
| Cytoskeleton | Microtubule | Smooth ER |
| Endoplasmic reticulum (ER) | Mitochondrion | Vacuole |
| Flagellum | Organelle | |

1. Any tiny structure that performs a specialized function in the cell _____

2. A plant organelle that may store starch or pigments _____

3. Converts the chemical energy in food into a form that is more easily used _____

4. Transport channels that are studded with ribosomes _____

5. Modifies, collects, packages, and distributes proteins that are produced by the cell _____

6. Saclike storage structure found in both animal and plant cells _____

7. Short threadlike structure that helps a unicellular organism move _____

8. Traps energy from sunlight and converts it to chemical energy _____

9. Contains digestive enzymes that help clean up the cell _____

10. Framework of filaments and fibers involved in cell support and movement _____

11. An organelle that serves as a protein factory _____

12. Tiny hollow tube made of proteins that is involved with support, the movement of organelles within the cell, and the formation of centrioles _____

13. Long, thin fiber that functions in the movement and support of the cell _____

14. A complex network of channels that is involved with transport, storage, and making and modifying proteins _____

15. Found only in plant and algae cells, it consists of two envelopelike membranes that surround a folded inner membrane _____

## Making Comparisons: Using the Main Ideas

1. How are mitochondria and chloroplasts similar?_____

_____

_____

_____

How are they different? _____

_____

_____

_____

2. How are lysosomes, vacuoles, and plastids similar?_____

_____

_____

_____

How are they different? _____

_____

_____

_____

## Concept Mapping

The construction of and theory behind concept mapping are discussed on pages vii–ix in the front of this Study Guide. Read those pages carefully. Then consider the concepts presented in Section 5–3 and how you would organize them into a concept map. Now look at the concept map for Chapter 5 on page 54. Notice that the concept map has been started for you. Add the key facts and concepts you feel are important for Section 5–3. When you have finished the chapter, you will have a completed concept map.

## SECTION REVIEW

In this section you learned about processes by which materials enter and leave the cell. These processes are diffusion, osmosis, facilitated diffusion, and active transport.

Diffusion is the process by which molecules of a substance move from areas of higher concentration to areas of lower concentration. Whether a substance will diffuse across a membrane depends on the permeability of the membrane to that substance as well as the concentration of the substance on either side of the membrane.

Osmosis is the diffusion of water molecules through a selectively permeable membrane.

Facilitated diffusion is the process in which carrier proteins help transport certain molecules across the cell membrane. This allows materials to be transported across the membrane quickly and selectively. But it can work only if the correct kind of concentration difference exists across the membrane.

Active transport enables materials to move across a cell membrane against a concentration difference. Active transport requires energy. There are two types of active transport. In one type, transport macromolecules pump materials across the membrane. In the other type, movements of the cell membrane take in or remove relatively large amounts of material through phagocytosis, pinocytosis, exocytosis, or the use of a contractile vacuole.

### Identifying Processes: Using the Main Ideas

Name the process involved in moving molecules across the cell membrane in each of the following cases. When appropriate, use a more specific term. (For example, "cat" would be a better answer than "animal.")

1. Fresh water moves into a single-celled organism.

   _____

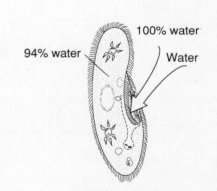

2. Pockets of the cell membrane fill with water and pinch off to become vacuoles inside the cell.

   _____

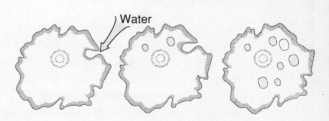

**3.** Oxygen molecules (O$_2$) move from the lungs into the bloodstream.

_____

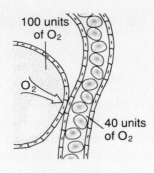

100 units
of O$_2$

O$_2$

40 units
of O$_2$

**4.** An ameba engulfs a large particle of food.

_____

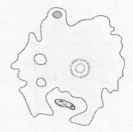

**5.** Carrier proteins transport glucose into a muscle cell.

_____

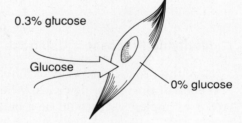

0.3% glucose

Glucose

0% glucose

**6.** Sodium ions (Na$^+$) are pumped out of a red blood cell.

_____

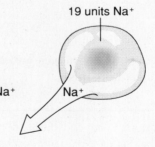

19 units Na$^+$

155 units Na$^+$

Na$^+$

### Concept Mapping

The construction of and theory behind concept mapping are discussed on pages vii–ix in the front of this Study Guide. Read those pages carefully. Then consider the concepts presented in Section 5–4 and how you would organize them into a concept map. Now look at the concept map for Chapter 5 on page 54. Notice that the concept map has been started for you. Add the key facts and concepts you feel are important for Section 5–4. When you have finished the chapter, you will have a completed concept map.

## Section 5–5 Cell Specification

### SECTION REVIEW

In this section you were introduced to the concept of cell specialization. You then read about three of the many kinds of specialized cells.

Recall that specialized cells are uniquely suited to perform a particular function within an organism. For example, some pancreas cells are specialized to produce large quantities of digestive enzymes. Certain eye cells have special structures that permit the cells to act as light detectors. And two kinds of cells help keep the air passageways in the body clean: The first kind of cell produces a sticky mucus that traps particles of dust and dirt and the second kind of cell has cilia that sweep dust, dirt, and excess mucus out of the passageways.

### Relating Structure and Function: Finding the Main Idea

In your own words, explain how each type of cell is specialized for its function. Use complete sentences.

**1.** Pancreas cells that produce digestive enzymes: _____

_____

_____

_____

**2.** Light-sensitive eye cells: _____

_____

_____

_____

_____

_____

**3.** Cells lining the air passageways:_____

_____

_____

### Concept Mapping

The construction of and theory behind concept mapping are discussed on pages vii–ix in the front of this Study Guide. Read those pages carefully. Then consider the concepts presented in Section 5–5 and how you would organize them into a concept map. Now look at the concept map for Chapter 5 on page 54. Notice that the concept map has been started for you. Add the key facts and concepts you feel are important for Section 5–5. When you have finished the chapter, you will have a completed concept map.

**Levels of Organization** *(pages 106–107)*

### SECTION REVIEW

One way to describe cells might be to list the function and position of every one of the cells in an organism. This would be more difficult than trying to describe the United States by making up a list of the names and occupations of its residents. And neither of these lists would be very useful or informative.

To effectively describe the United States, you might divide its residents into groups using levels of organization such as states, counties, towns, and individual citizens. To effectively describe organisms, biologists also use levels of organization.

Remember that the organization of the body's cells into tissues, organs, and organ systems makes possible a division of labor among those cells. This allows for specialization, which leads to interdependence.

### Fill in the Blanks: Building Vocabulary Skills

In the spaces provided, write in the correct terms.

The first level of organization is the _____. Organisms that possess only the first level of organization include amebas, bacteria, yeasts, and other unicellular organisms.

The second level of organization is the _____. A (An) _____ is a group of similar cells that perform similar functions. The cells that line your air passages are one kind of _____. So are the cells that respond to light in the retina and the cells that produce digestive enzymes in the pancreas.

The third level of organization is the _____. Although hundreds of thousands of _____ may be involved in forming a tissue, many tasks are too complex to be carried out by a single tissue. A (An) _____ is a group of tissues put together to perform a specific function. Many types of tissues may be used to form a particular _____. Each of these _____ performs an essential task to help the _____ function successfully.

The fourth level of organization is the _____. In many cases, even a complex _____ is not sufficient to complete a series of specialized tasks. A (An) _____ is a group of organs that perform related tasks.

### Concept Mapping

The construction of and theory behind concept mapping are discussed on pages vii–ix in the front of this Study Guide. Read those pages carefully. Then consider the concepts presented in Section 5–6 and how you would organize them into a concept map. Now look at the concept map for Chapter 5 on page 54. Notice that the concept map has been started for you. Add the key facts and concepts you feel are important for Section 5–6. When you have finished the chapter, you will have a completed concept map.

## Using the Writing Process

Use your writing skills and imagination to respond to the following writing assignment. You will probably need an additional sheet of paper to complete your response.

Imagine that the cell is a small town. The various cell parts and the organelles are the residents of the community. These residents all have their own unique job and personality. For example, the protective and regulating cell membrane might be a police officer—or somebody's overbearing parent. Write a short story or play that introduces the members of this town. Be sure to name the town. Be creative and have a good time.

_____

_____

_____

_____

_____

_____

_____

_____

_____

_____

_____

_____

_____

_____

_____

_____

_____

_____

_____

_____

## Concept Mapping

The concept map below has been started for you. Add the key facts and concepts for each section of the chapter to this partial concept map. When you are done, you will have a concept map for the entire chapter.

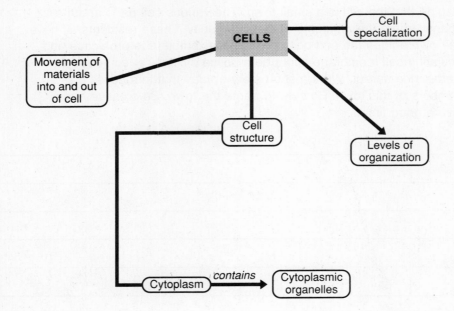

STUDY
GUIDE

**Section 6–1**  **Photosynthesis: Capturing and Converting Energy**  *(pages 113–117)*

### SECTION REVIEW

In this section you were introduced to the process of photosynthesis. Photosynthesis is the process by which the energy of sunlight is converted into the energy in the chemical bonds of carbohydrates. Photosynthesis is usually associated with green plants, although it also occurs in a number of microorganisms.

In the first part of this section you read about some early studies that were essential to the development of our understanding of photosynthesis. The experiments performed by Van Helmont, Priestley, Ingenhousz, and other scientists demonstrated that in the presence of light, plants transform water and carbon dioxide into carbohydrates and release oxygen. Because the carbohydrate produced by photosynthesis is often glucose, you can summarize the process of photosynthesis with the following chemical equation:

$$6\ CO_2 + 6\ H_2O \xrightarrow{\text{light}} C_6H_{12}O_6 + 6\ O_2.$$

In the second part of this section you learned about some of the requirements of photosynthesis. Recall that sunlight, pigments such as chlorophyll, and energy-storing compounds such as ATP are as necessary for photosynthesis as carbon dioxide and water. You also learned that organisms can be separated into two groups according to their methods of obtaining energy. Autotrophs are organisms that are able to use a source of energy, such as sunlight, to produce food directly from simple inorganic molecules in the environment. Heterotrophs cannot make their own food. Instead, they obtain the energy they need to live from the foods they eat.

### What—or Who—is It?: Building Vocabulary Skills

In the space provided, write the name of the scientist or the vocabulary term that best fits each description.

**1.** The light-requiring process in which carbon dioxide and water combine to form

carbohydrates and oxygen gas: _____

**2.** An organism that is able to use a source of energy to produce food from simple

inorganic molecules: _____

**3.** Performed an experiment that showed that plants produce a gas that allows a candle to

burn: _____

**4.** The most important energy-storing compound: _____

**5.** An organism that obtains energy by eating food: _____

**6.** A green pigment that is involved in light absorption for photosynthesis:

_____

### Autotroph or Heterotroph?: Applying Concepts

In the space beneath each picture, identify the organism shown as either an autotroph or a heterotroph.

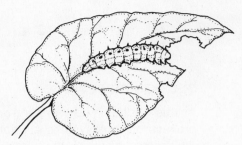

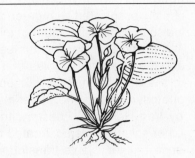

### Understanding Concepts: Finding the Main Ideas

1. In your own words, define the term *photosynthesis*.

_____

_____

_____

2. Explain how ATP, ADP, and AMP store and release energy. _____

_____

_____

_____

### Concept Mapping

The construction of and theory behind concept mapping are discussed on pages vii–ix in the front of this Study Guide. Read those pages carefully. Then consider the concepts presented in Section 6–1 and how you would organize them into a concept map. Now look at the concept map for Chapter 6 on page 66. Notice that the concept map has been started for you. Add the key facts and concepts you feel are important for Section 6–1. When you have finished the chapter, you will have a completed concept map.

## Section 6–2   Photosynthesis: The Light and Dark Reactions   *(pages 118–123)*

### SECTION REVIEW

Recall that photosynthesis is divided into two parts: the light reactions and the dark reactions. Both sets of reactions occur in the chloroplast. The light reactions occur in the photosynthetic membranes of the chloroplast. The dark reactions occur in the areas surrounding the photosynthetic membranes. Remember that the dark reactions can, and usually do, occur in sunlight. However, light does not play a role in the dark reactions.

In the light reactions, the energy of sunlight is captured and used to make energy-storing compounds. The light reactions can be divided into four basic processes: light absorption, electron transport, oxygen production, and ATP formation. The light reactions use water, ADP, and $NADP^+$. They produce oxygen ($O_2$) and the energy-storing molecules ADP and NADPH.

In the dark reactions, carbon dioxide is used to make a complex organic molecule called PGAL via a circular series of reactions. Energy for the dark reactions is supplied by ATP and NADPH, which are made by the light reactions. The PGAL produced in the dark reactions is used to make biologically important molecules such as glucose. Because the chemistry of this part of photosynthesis was worked out by Melvin Calvin, the dark reactions are also known as the Calvin cycle. Although the production of glucose is usually emphasized when the dark reactions are studied, it is important to remember that the intermediate compounds in the cycle can be used to make amino acids, lipids, or sugars other than glucose. Thus the dark reactions provide the raw material to produce almost every organic compound needed by the cell.

### Fill in the Blanks: Building Vocabulary Skills

Complete each of the following sentences by writing in the correct terms.

1. The _____ take place within the saclike _____ that

   are located inside an organelle known as the _____.

2. Sunlight is captured by clusters of pigment molecules called _____ that

   contain several hundred _____ molecules as well as a number of
   accessory pigments.

3. Electrons are passed from one _____ to the next during the process of

   _____.

4. _____ is "split" to produce hydrogen ions, _____ gas, and

   _____ that replace those lost by chlorophyll.

5. An enzyme uses the energy created by a difference in charges across a membrane to add

   a (an) _____ group to ADP to form _____.

**6.** The light reactions use _____ , _____ , and

_____ and produce _____ , _____ , and

_____ .

**7.** In the _____ , or Calvin cycle, the energy from the

_____ and _____ produced in the light reactions is

used to make _____ into PGAL and other biologically important
molecules.

**8.** In the first reaction of the Calvin cycle, which is catalyzed by the enzyme

_____ , a 5-carbon sugar combines with _____ .

### Making Diagrams: Using the Main Ideas

Draw a labeled diagram of the process of photosynthesis in the space
provided. Your diagram should be simpler than the one in your text. It is
not necessary to show the structure of the photosynthetic membrane.

### Concept Mapping

The construction of and theory behind concept mapping are discussed on
pages vii–ix in the front of this Study Guide. Read those pages carefully. Then
consider the concepts presented in Section 6–2 and how you would organize
them into a concept map. Now look at the concept map for Chapter 6 on
page 66. Notice that the concept map has been started for you. Add the key
facts and concepts you feel are important for Section 6–2. When you have
finished the chapter, you will have a completed concept map.

## Section 6-3 Glycolysis and Respiration

*(pages 123–129)*

### SECTION REVIEW

Remember that the energy captured by photosynthesis is generally stored in the chemical bonds of the sugar glucose. This energy is released when the chemical bonds of glucose are broken. In this section you learned about two processes that release energy from glucose: glycolysis and respiration.

Glycolysis, which takes place in the cytoplasm of a cell, is the first stage in the breakdown of glucose. Glycolysis is a series of enzyme-mediated chemical reactions that change glucose one step at a time into different molecules. Ultimately, glycolysis transforms 1 molecule of 6-carbon glucose into 2 molecules of 3-carbon pyruvic acid. It also produces a small amount of energy in the form of ATP and NADH.

Respiration continues the breakdown of glucose by breaking down pyruvic acid to produce 34 molecules of ATP. In eukaryotes, respiration takes place in the mitochondria of a cell. The process of respiration can be divided into two parts: the Krebs cycle and the electron-transport chain.

Respiration can be defined as an oxygen-requiring process in which food molecules are broken down to release energy. Remember that the oxygen needed for respiration is taken in by breathing. And the food molecules used in respiration are not limited to glucose.

When a glucose molecule is totally broken down in the presence of oxygen, the energy produced is equivalent to 36 molecules of ATP. Following is the equation for this reaction: $C_6H_{12}O_6 + 6 O_2 \rightarrow 6 CO_2 + 6 H_2O + energy$. Note that this is the reverse of the equation for photosynthesis:
$6 CO_2 + 6 H_2O + energy \rightarrow C_6H_{12}O_6 + 6 O_2$.

### Relating Terms: Building Vocabulary Skills

In your own words, explain how the paired terms are related to each other.

**1.** Krebs cycle, electron-transport chain _____

_____

_____

_____

_____

_____

_____

**2.** PGAL, pyruvic acid _____

_____

_____

_____

_____

_____

**3.** Glycolysis, respiration _____

_____

_____

_____

_____

### Examining Glycolysis and Respiration: Finding the Main Ideas

**1.** Define the term *calorie*. _____

_____

_____

_____

_____

How many calories are released when 1 gram of glucose is completely broken down in

the presence of oxygen? _____

How does this amount of energy compare to that released by glycolysis and respiration?

_____

_____

_____

_____

_____

_____

_____

**2.** What is the net energy gain in glycolysis? Explain. _____

_____

_____

_____

_____

_____

**3.** What happens to the NADH produced in glycolysis? _____

_____

_____

What happens to the pyruvic acid produced in glycolysis? _____

_____

_____

_____

**4.** What happens to pyruvic acid before it enters the Krebs cycle?_____

_____

_____

_____

_____

**5.** What compounds are produced by the Krebs cycle for each molecule of acetic acid that

enters the cycle? _____

_____

_____

_____

What happens to the carbon dioxide produced in your body during respiration?

_____

_____

What happens to the NADH and $FADH_2$ molecules produced by the Krebs cycle?

_____

_____

_____

**6.** What happens to the high-energy electrons after they are passed to the electron-

transport chain?_____

_____

_____

_____

As the high-energy electrons move through the electron-transport chain, some of their energy is released. How is this energy used? _____

_____

_____

_____

**7.** Why is the difference in charge across the inner mitochondrial membrane important?

_____

_____

**8.** What happens to the high-energy electrons after they are passed to the electron-transport chain?_____

_____

_____

_____

Write a balanced equation for this reaction. _____

_____

Explain why respiration requires oxygen. _____

_____

_____

_____

How does your body obtain the oxygen needed for respiration? _____

_____

_____

_____

## Concept Mapping

The construction of and theory behind concept mapping are discussed on pages vii–ix in the front of this Study Guide. Read those pages carefully. Then consider the concepts presented in Section 6–3 and how you would organize them into a concept map. Now look at the concept map for Chapter 6 on page 66. Notice that the concept map has been started for you. Add the key facts and concepts you feel are important for Section 6–3. When you have finished the chapter, you will have a completed concept map.

| Section 6–4 | Fermentation | (pages 130–131) |
|---|---|---|

## SECTION REVIEW

How is energy obtained from glucose when oxygen is not present? In this section you discovered the answer to this question as you were introduced to the process of fermentation. Fermentation is an anaerobic process that enables cells to carry out energy production in the absence of oxygen. The combination of glycolysis, which is also anaerobic, and fermentation produces 2 molecules of ATP from a molecule of glucose.

Recall that glycolysis produces 2 molecules of pyruvic acid, 2 molecules of NADH, and a total of 2 molecules of ATP. In order for glycolysis to continue, the NADH must be converted back to $NAD^+$. In fermentation, electrons from NADH are transferred to an organic molecule such as pyruvic acid. This converts the NADH to $NAD^+$, which can then be used in glycolysis.

In this section you learned that most eukaryotic cells use one of two fermentation pathways: lactic acid fermentation or alcoholic fermentation. In cells that undergo lactic acid fermentation—human muscle cells, for example—electrons from NADH are added to pyruvic acid, transforming it into lactic acid. In cells that undergo alcoholic fermentation—yeast cells, for example—electrons from NADH are added to pyruvic acid, changing it into alcohol and carbon dioxide.

### Defining Terms: Building Vocabulary Skills

In your own words, define each of the following terms.

**1.** Anaerobic: _____

_____

**2.** Aerobic: _____

_____

**3.** Fermentation: _____

_____

_____

**4.** Lactic acid fermentation: _____

_____

_____

**5.** Alcoholic fermentation: _____

_____

_____

**6.** Pyruvic acid: _____

_____

### Matching Up: Using the Main Ideas

In the space provided, write the term from the following word bank that best fits the description. Some terms are used more than once.

| Word Bank | |
|---|---|
| Respiration | Lactic acid fermentation |
| Glycolysis | Alcoholic fermentation |

**1.** Important to bakers and brewers: _____

**2.** A great deal of energy is extracted from glucose: _____

**3.** Causes a painful, burning sensation in muscles: _____

**4.** Produces 2 ATP molecules and pyruvic acid: _____

**5.** Associated with rapid exercise: _____

**6.** Fermentation that occurs in yeast cells: _____

**7.** Requires oxygen: _____

**8.** Fermentation that occurs in muscle cells that do not have sufficient oxygen:

_____

**9.** Alcohol is a by-product: _____

**10.** Pyruvic acid is the final electron acceptor (2 answers): _____

_____

### Concept Mapping

The construction of and theory behind concept mapping are discussed on pages vii–ix in the front of this Study Guide. Read those pages carefully. Then consider the concepts presented in Section 6–4 and how you would organize them into a concept map. Now look at the concept map for Chapter 6 on page 66. Notice that the concept map has been started for you. Add the key facts and concepts you feel are important for Section 6–4. When you have finished the chapter, you will have a completed concept map.

## Using the Writing Process

Use your writing skills and imagination to respond to the following writing assignment. You will probably need an additional sheet of paper to complete your response.

It has been said many times that to really understand something, you should teach it to someone else. Pretend that you have been selected to teach first grade students about the processes that involve energy transformations in living organisms. As you know, this subject can be complicated and even boring, so use all of your powers of imagination to make it interesting. Explain the topic of cell energy in the form of a fairy tale. By the way, it is a rule that fairy tales begin with "Once upon a time" and end with "and they lived happily ever after."

_____

_____

_____

_____

_____

_____

_____

_____

_____

_____

_____

_____

_____

_____

_____

_____

_____

## Concept Mapping

The concept map below has been started for you. Add the key facts and concepts for each section of the chapter to this partial concept map. When you are done, you will have a concept map for the entire chapter.

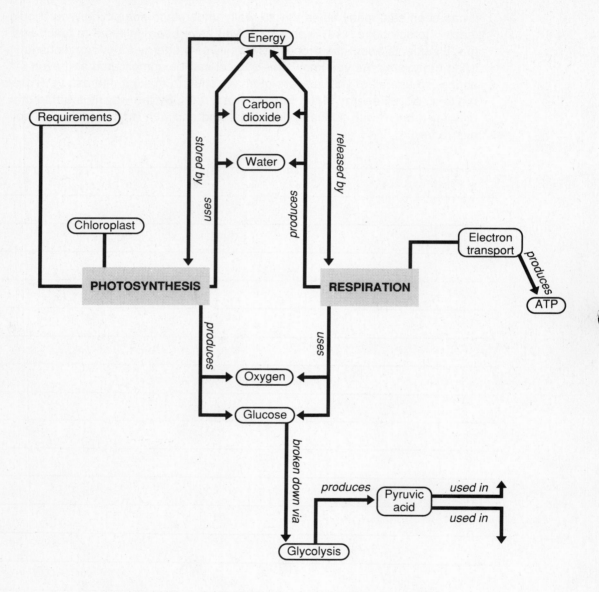

S T U D Y
G U I D E

**Section 7–1**   **Protein Synthesis**                                    *(pages 137–145)*

### SECTION REVIEW

In this section you were introduced to the concept of the genetic code and the molecule that carries the genetic code: DNA. First, you read about the experiments that demonstrated that DNA was the crucial molecule for the passing on of genetic information. Griffith's experiments with pneumonia bacteria showed that heat-killed bacteria could pass their disease-causing ability to harmless live bacteria. The process of transformation was further studied by a team of scientists led by Avery. Avery's team found that transformation did not occur when DNA was destroyed. This indicated that DNA was the transforming factor. The conclusions of Avery's team were confirmed by an experiment performed by Hershey and Chase. Using viruses labeled with radioactive sulfur or with radioactive phosphorus, Hershey and Chase demonstrated that viruses inject phos-

phorus-containing DNA into bacteria cells. It is this injected DNA that alters the genetic information inside an infected cell and causes virus particles to be made.

Next, you found out that DNA is composed of a string of nucleotides. Each nucleotide consists of three parts: a sugar called deoxyribose, a phosphate group, and a nitrogenous base. You also learned a little about the scientists who unraveled the mystery of DNA structure. Recall that data from Franklin and Wilkins's X-ray diffraction studies and from Chargaff's biochemical studies gave Watson and Crick clues about the structure of DNA. These clues helped them develop their double-helix model of DNA.

Finally, you learned about the process of DNA replication. Replication is the process by which a cell makes an exact copy of its DNA.

### Fill in the Blanks: Building Vocabulary Skills

Using what you have learned in Section 7–1, complete each of the following sentences.

1. Each nucleotide in DNA consists of three basic parts: a sugar called

   _____, a _____ group, and a _____ .

2. The _____ gives cells information about what to do and how to do it.

3. In one example of the process of _____, heat-killed bacteria transfer their disease-causing ability to harmless live cells.

4. A team of scientists led by _____ determined that the transforming factor

   was _____ .

5. DNA is duplicated in the process of _____ .

6. _____ and _____ worked with _____, which are viruses that infect bacteria cells.

7. The individual units in DNA are known as _____.

8. _____ and _____ performed X-ray diffraction studies of DNA.

9. In DNA, adenine pairs with _____, whereas _____ pairs with guanine.

10. _____ observed the process of transformation when he was studying _____.

11. The double-helix model of DNA was proposed by _____ and _____ .

12. _____ is the force that holds the two strands of the DNA molecule together.

### Applying Concepts: Using the Main Ideas

Place the diagrams in the correct order by writing the numbers 1 through 3 above each diagram. Then answer the questions that follow.

_____     _____     _____

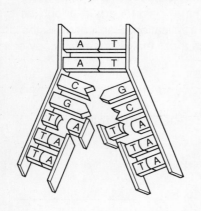
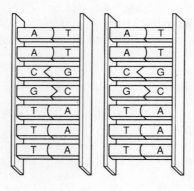
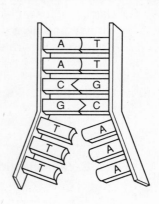

Complementary bases attach to template.

Two molecules of DNA identical to each other and the original molecule are made.

DNA "unzips."

1. What process is shown in these diagrams? _____

_____

**2.** How are enzymes involved in this process? _____

_____

_____

_____

_____

**3.** What happens when DNA "unzips"? _____

_____

_____

_____

_____

**4.** Why is it important that exact copies of DNA be made? _____

_____

_____

_____

_____

**5.** Suppose that a sequence of one DNA strand is T-A-C-A-A-C-G-T-G. What is the

corresponding sequence on the other strand? _____

_____

_____

_____

_____

### Concept Mapping

The construction of and theory behind concept mapping are discussed on pages vii–ix in the front of this Study Guide. Read those pages carefully. Then consider the concepts presented in Section 7–1 and how you would organize them into a concept map. Now look at the concept map for Chapter 7 on page 74. Notice that the concept map has been started for you. Add the key facts and concepts you feel are important for Section 7–1. When you have finished the chapter, you will have a completed concept map.

| Section 7–2 | **RNA** | *(pages 146–148)* |

## SECTION REVIEW

In this section you were introduced to the molecule that helps put the information in DNA to use: ribonucleic acid, or RNA. RNA is a nucleic acid that carries information from DNA to the ribosomes, the organelles in which proteins are made. RNA also carries out the process by which proteins are assembled from amino acids.

RNA is quite similar to DNA in structure. However, there are some important differences. RNA is single-stranded; DNA is double-stranded. RNA contains the sugar ribose; DNA contains deoxyribose. And RNA has the nitrogenous base uracil; DNA has thymine.

In this section you also learned about the process of transcription. Transcription is the process in which part of a DNA molecule is used as a template for the synthesis of a complementary strand of RNA. This process is mediated by an enzyme called RNA polymerase. The strand of RNA formed by transcription is called mRNA. Its function is to carry the genetic information from the DNA in the nucleus to the ribosomes in the cytoplasm.

## Comparing RNA and DNA: Finding the Main Ideas

Carefully read each term or phrase in the left-hand column of the following table. If the term or phrase applies to DNA, place a check mark in the column labeled DNA. If it applies to RNA, place a check mark in the column labeled RNA. If it applies to both nucleic acids, place a check mark in both columns.

| | DNA | RNA | | DNA | RNA |
|---|---|---|---|---|---|
| Nucleotides | | | Double helix | | |
| Deoxyribose | | | Replication | | |
| Ribose | | | Transcription | | |
| Single-stranded | | | Exact copy | | |
| Double-stranded | | | Messenger | | |
| Nitrogenous bases | | | More than one form | | |
| Thymine | | | Found in nucleus | | |
| Uracil | | | Leaves nucleus | | |
| Template for synthesis of nucleic acid | | | Does not leave nucleus | | |

## Concept Mapping

The construction of and theory behind concept mapping are discussed on pages vii–ix in the front of this Study Guide. Read those pages carefully. Then consider the concepts presented in Section 7–2 and how you would organize them into a concept map. Now look at the concept map for Chapter 7 on page 74. Notice that the concept map has been started for you. Add the key facts and concepts you feel are important for Section 7–2. When you have finished the chapter, you will have a completed concept map.

| Section 7–3 | Protein Synthesis | *(pages 148–153)* |

## SECTION REVIEW

In this section you studied the process of protein synthesis. You learned that the information that DNA transfers to messenger RNA (mRNA) is in the form of a code. When the information is decoded, chains of amino acids, called polypeptides, are formed. Polypeptides make up proteins, which direct biochemical pathways and are responsible for cell structure and movement.

The genetic code is determined by the arrangement of the nitrogenous bases in DNA and RNA. A code word in DNA consists of a group of three nucleotides. When transcribed into mRNA, each code word, or codon, designates a specific amino acid that is to be placed in the polypeptide chain. More than one codon may code for a particular amino acid. The mRNA sequence AUG serves as an initiator, or "start," codon. Three other sequences serve as "stop" codons.

In the second part of this section you learned about the process of translation. During translation, each mRNA codon in turn matches up with the complementary transfer RNA (tRNA) anticodon. As this occurs, the amino acid that is specified by the codon and carried by the tRNA is added to the polypeptide chain being formed. Once an amino acid is added to the polypeptide, the tRNA molecule that carried it is free to pick up another amino acid. Translation ends when the ribosome reaches a stop codon on the mRNA.

### Describing Functions: Building Vocabulary Skills

In your own words, define each of the following in terms of its function.

**1.** Codon: _____

_____

_____

**2.** Anticodon: _____

_____

_____

**3.** Transfer RNA (tRNA): _____

_____

_____

**4.** Ribosomal RNA (rRNA): _____

_____

_____

**5.** Translation: _____

_____

_____

### Interpreting Diagrams: Understanding the Main Ideas

#### The Genetic Code (mRNA)

| First Base in Code Word | | Second Base in Code Word | | | | Third Base in Code Word |
|---|---|---|---|---|---|---|
| | | A | G | U | C | |
| A | | Lysine<br>Lysine<br>Asparagine<br>Asparagine | Arginine<br>Arginine<br>Serine<br>Serine | Isoleucine<br>Methionine<br>Isoleucine<br>Isoleucine | Threonine<br>Threonine<br>Threonine<br>Threonine | A<br>G<br>U<br>C |
| G | | Glutamic acid<br>Glutamic acid<br>Aspartic acid<br>Aspartic acid | Glycine<br>Glycine<br>Glycine<br>Glycine | Valine<br>Valine<br>Valine<br>Valine | Alanine<br>Alanine<br>Alanine<br>Alanine | A<br>G<br>U<br>C |
| U | | "Stop" codon<br>"Stop" codon<br>Tyrosine<br>Tyrosine | "Stop" codon<br>Trytophan<br>Cysteine<br>Cysteine | Leucine<br>Leucine<br>Phenylalanine<br>Phenylalanine | Serine<br>Serine<br>Serine<br>Serine | A<br>G<br>U<br>C |
| C | | Glutamine<br>Glutamine<br>Histidine<br>Histidine | Arginine<br>Arginine<br>Arginine<br>Arginine | Leucine<br>Leucine<br>Leucine<br>Leucine | Proline<br>Proline<br>Proline<br>Proline | A<br>G<br>U<br>C |

Use the information in the accompanying figure to complete the following table. The first row has been completed to help you get started.

| DNA codon | mRNA codon | tRNA Anticodon | Amino Acid |
|---|---|---|---|
| AAA | UUU | AAA | Phenylalanine |
| GTC | | | |
| | GGA | | "Stop" |
| | | | Methionine/"Initiator" |
| GAT | | | |
| | GUG | | Valine |

### Concept Mapping

The construction of and theory behind concept mapping are discussed on pages vii–ix in the front of this Study Guide. Read those pages carefully. Then consider the concepts presented in Section 7–3 and how you would organize them into a concept map. Now look at the concept map for Chapter 7 on page 74. Notice that the concept map has been started for you. Add the key facts and concepts you feel are important for Section 7–3. When you have finished the chapter, you will have a completed concept map.

## Using the Writing Process                     *Chapter 7*

Use your writing skills and imagination to respond to the following writing assignment. You will probably need an additional sheet of paper to complete your response.

Use your writing skills to explain the process of protein synthesis, from DNA replication to mRNA synthesis to formation of a polypeptide. Use your imagination to make the explanation more interesting. Treat the topic as though the entire process had just been discovered and you are telling the world about

## Interpreting Diagrams: Understanding the Main Ideas

### The Genetic Code (mRNA)

**Third Base in Code Word**

| First Base | Second Base: U | Second Base: C | Second Base: A | Second Base: G | Third Base |
|---|---|---|---|---|---|
| **U** | Phenylalanine / Phenylalanine / Leucine / Leucine | Serine / Serine / Serine / Serine | Tyrosine / Tyrosine / "Stop" codon / "Stop" codon | Cysteine / Cysteine / "Stop" codon / Tryptophan | U / C / A / G |
| **C** | Leucine / Leucine / Leucine / Leucine | Proline / Proline / Proline / Proline | Histidine / Histidine / Glutamine / Glutamine | Arginine / Arginine / Arginine / Arginine | U / C / A / G |
| **A** | Isoleucine / Isoleucine / Isoleucine / Methionine | Threonine / Threonine / Threonine / Threonine | Asparagine / Asparagine / Lysine / Lysine | Serine / Serine / Arginine / Arginine | U / C / A / G |
| **G** | Valine / Valine / Valine / Valine | Alanine / Alanine / Alanine / Alanine | Aspartic acid / Aspartic acid / Glutamic acid / Glutamic acid | Glycine / Glycine / Glycine / Glycine | U / C / A / G |

**First Base in Code Word**

**Second Base in Code Word**

## Concept Mapping

The concept map has been started for you. Add the key facts and concepts for each section of the chapter to this partial concept map. When you are done, you will have a concept map for the entire chapter.

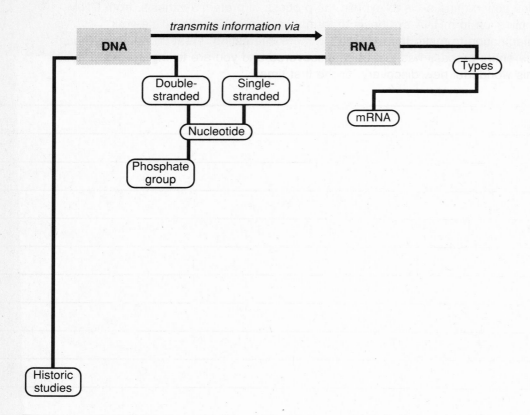

Name

## STUDY
## GUIDE

*Cell Growth a*

| Section 8–1 | **Cell Growth** | *(pages 15_* |
|---|---|---|

### SECTION REVIEW

In this section you examined the nature of cell growth. You learned that in most cases a living thing grows because it produces more cells, not because its cells get larger. Recall that cell size is limited for two reasons. First, because volume increases at a faster rate than surface area, a larger cell has more difficulty getting food and nutrients in and wastes out than a smaller cell. Second, at some point the DNA in a large cell may no longer be able to make enough RNA to supply the increasing needs of the growing cell. The problems of a too-large cell are solved if the cell is divided to form smaller, more efficient cells.

Under ideal conditions, cell division can produce large numbers of cells in a surprisingly short time. However, real conditions are seldom ideal. In addition, cell growth within an organism is usually controlled by mechanisms that are not yet understood. For example, some types of cells rarely divide. Others stop growing and dividing when they come into contact with other cells. Occasionally, some of the cells within a multicellular organism lose the ability to control their rate of growth. The uncontrolled growth of such cells causes the disorder known as cancer.

### Relating Concepts: Finding the Main Ideas

**1.** In your own words, define the terms *surface area* and *volume*.

_____

_____

_____

_____

_____

_____

**2.** What happens to the surface-to-volume ratio as the size of a cell increases? Why is this

significant? _____

_____

_____

_____

_____

_____

**3.** What is cell division? Why is it important? _____

_____

_____

_____

_____

_____

_____

_____

_____

**4.** What is cancer? Give two reasons for studying cancer. _____

_____

_____

_____

_____

_____

_____

_____

_____

_____

## Concept Mapping

The construction of and theory behind concept mapping are discussed on pages vii–ix in the front of this Study Guide. Read those pages carefully. Then consider the concepts presented in Section 8–1 and how you would organize them into a concept map. Now look at the concept map for Chapter 8 on page 82. Notice that the concept map has been started for you. Add the key facts and concepts you feel are important for Section 8–1. When you have finished the chapter, you will have a completed concept map.

| Section 8–2 | **Cell Division: Mitosis and Cytokinesis** | *(pages 164–171)* |

## SECTION REVIEW

In this section you examined the process by which a eukaryotic cell divides to form two daughter cells that are genetically identical to one another and to the original cell. This process of cell division, which is sometimes informally called mitosis after one of its stages, is a brief but interesting part of the cell cycle.

The cell cycle is the period from the beginning of one period of cell division to the beginning of the next. During the first, or $G_1$, phase of a cell cycle, a cell grows and develops. In the next two phases, the cell prepares for cell division. During the S phase, the cell replicates its DNA and produces certain proteins associated with the chromosomes. During the $G_2$ phase, the cell synthesizes organelles and materials required for cell division. During the fourth and final (M) phase, the cell divides to form two daughter cells. The $G_1$,

S, and $G_2$ phases are collectively known as interphase. The M phase is divided into two stages: mitosis and cytokinesis.

Mitosis is the process by which the nucleus of the cell is divided into two nuclei, each of which has the same number and kinds of chromosomes as the parent cell. Mitosis is divided into four phases; prophase, metaphase, anaphase, and telophase. Recall that certain chromosome movements and other cellular activities are associated with each phase.

Cytokinesis quickly follows mitosis. Cytokinesis is the process by which the cytoplasm of the cell divides to form two individual cells. In animal cells, cytokinesis is accomplished through a constriction of the cell membrane. In plant cells, it is accomplished by the formation of a cell plate.

### Writing Definitions: Building Vocabulary Skills

For each of the following pairs of terms, define the first term using the second term in your definition.

**1.** Mitosis, chromosome: _____

_____

_____

_____

_____

**2.** Chromosome, chromatin: _____

_____

_____

_____

_____

**3.** Chromatids, chromosomes: _____

_____

_____

_____

_____

**4.** Centromere, chromatid: _____

_____

_____

_____

_____

**5.** Cell cycle, daughter cells: _____

_____

_____

_____

_____

**6.** Interphase, mitosis: _____

_____

_____

_____

_____

**7.** Spindle, centriole: _____

_____

_____

_____

_____

**8.** Cytokinesis, cytoplasm: _____

_____

_____

_____

_____

### What's in a Name?: Analyzing Terms

The names that were coined to describe the parts of the cell cycle seemed quite appropriate and descriptive at first. However, as our understanding of the cell cycle improved, many of these names no longer seemed to fit as well. Some even began to seem misleading. Consider each of the following terms. Do you think the term is appropriate, inappropriate, or a little of both? Explain your reasons. *Hint:* You may find it useful to consult a dictionary to examine the derivation of some of the terms.

1. $G_2$ (Gap 2): _____

_____

_____

_____

2. M (Mitosis): _____

_____

_____

3. Interphase: _____

_____

_____

### Analyzing Mitosis: Understanding the Main Ideas

Write the name of the phase beneath each of the following diagrams. Then use the diagrams and what you learned in Section 8–2 to complete the table on the following page.

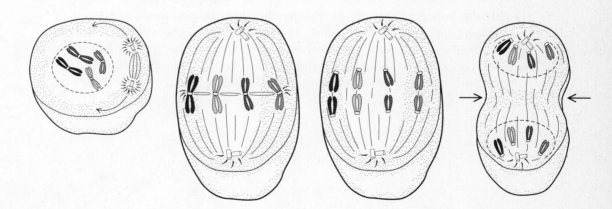

_____  _____  _____  _____

| Phase | Begins When | Events During | | Ends When |
| | | Chromosomes | Other | |
| --- | --- | --- | --- | --- |
| | Chromosomes become visible | | | |
| | | Line up along equator | | |
| | | | Chromosomes moved by unknown mechanism | Movement of chromosome stops |
| Telophase | | | | |

■ **Concept Mapping**

The construction of and theory behind concept mapping are discussed on pages vii–ix in the front of this Study Guide. Read those pages carefully. Then consider the concepts presented in Section 8–2 and how you would organize them into a concept map. Now look at the concept map for Chapter 8 on page 82. Notice that the concept map has been started for you. Add the key facts and concepts you feel are important for Section 8–2. When you have finished the chapter, you will have a completed concept map.

## Using the Writing Process                              *Chapter 8*

Use your writing skills and imagination to respond to the following writing assignment. You will probably need an additional sheet of paper to complete your response.

Imagine that you are a chromosome. Describe the events of the cell cycle from your new point of view. This description may be in the form of a diary, an essay, or a short story.

*Hint:* Before you begin, you might wish to consider the following questions. What sorts of structures would a chromosome encounter? What kind of personality might these structures have? How might a chromosome respond to acquiring a "Siamese twin" during the S phase of the cell cycle? How would the chromosome feel about being separated from this twin during mitosis?

_____

_____

_____

_____

_____

_____

_____

_____

_____

_____

_____

_____

_____

_____

_____

_____

_____

_____

## Concept Mapping

## Chapter 8

The concept map below has been started for you. Add the key facts and concepts for each section of the chapter to this partial concept map. When you are done, you will have a concept map for the entire chapter.

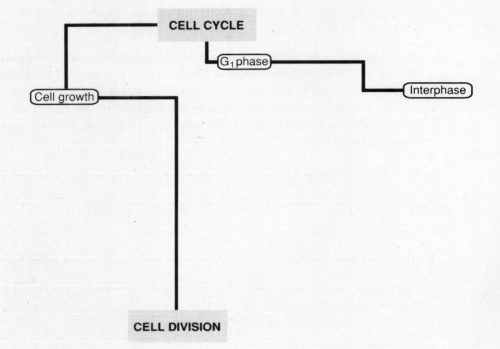

S T U D Y
G U I D E

CHAPTER
9
Introduction to Genetics

**Section 9–1**    **The Work of Gregor Mendel**    *(pages 181–189)*

### SECTION REVIEW

In this section you read about Gregor Mendel's experiments with pea plants. You also were introduced to the Punnett square, a diagram used to show crosses in genetics. Recall that Mendel analyzed the data from his experiments using mathematical principles. This innovation in data analysis enabled him to see the basic patterns of heredity. Although his work remained unappreciated during his lifetime, Mendel is now recognized for making an important breakthrough in biology: correctly describing the basic principles of genetics.

Recall that four key terms are associated with Mendel's work: genes, dominance, segregation, and independent assortment. Genes are individual factors that control each trait of a living thing. In organisms that reproduce sexu-

ally, genes are inherited from each parent. The traits that Mendel studied were controlled by one gene that occurred in two contrasting forms, or alleles. Dominance has to do with which allele's effects are seen when both are present in an organism. The effects of a dominant allele are seen even when the contrasting recessive allele is present. The effects of a recessive allele are seen only when the contrasting dominant allele is absent. Segregation refers to the fact that paired alleles are segregated, or separated, during gamete formation. And independent assortment refers to the fact that the genes for different traits may segregate independently of one another. This produces new combinations of traits in the gametes and offspring.

### Contrasting Terms: Building Vocabulary Skills

In the space provided, explain how the paired terms are different from each other.

**1.** Self-pollination, cross-pollination: _____

_____

_____

_____

_____

**2.** Purebred, hybrid: _____

_____

_____

_____

_____

**3.** Gene, alleles: _____

_____

_____

_____

_____

**4.** Dominant, recessive: _____

_____

_____

_____

_____

**5.** Segregation, independent assortment: _____

_____

_____

_____

_____

**6.** Heredity, genetics: _____

_____

_____

_____

_____

**7.** Phenotype, genotype: _____

_____

_____

_____

_____

**8.** Homozygous, heterozygous: _____

_____

_____

_____

_____

### ▨ Segregation of Alleles: Applying the Main Ideas

Complete Figure 1 as follows.

1.  Select one of the pea traits shown in Figure 9–3 on page 183 in your textbook. Choose a letter to represent that trait. (*Hint:* Use a letter whose capital form does not look too similar to its lowercase form.) Then fill in the circles by writing in the allele(s) found in the individuals and in the gametes.

2.  Label Figure 1 using the following terms: $F_1$, $F_2$, *Gametes*, *P*, *Segregation*, *Fertilization*. **Note:** Some of these terms may be used more than once.

3.  Complete the key for Figure 1 that appears on the following page. If one of the column heads does not apply to a circle, put an X in the appropriate place in the key.

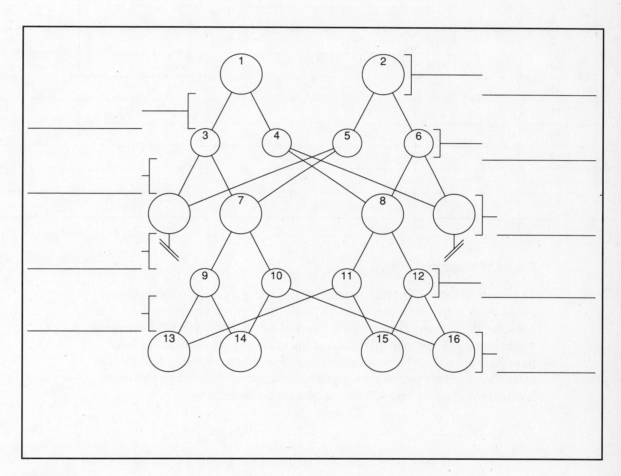

**Figure 1**

| KEY | | | | | |
|---|---|---|---|---|---|
| Circle Number | Genotype | Heterozygous or Homozygous? | Phenotype | Generation | Gamete? |
| 1 | | Homozygous | | | |
| 2 | | Homozygous | | | |
| 3 | | X | X | X | Yes |
| 4 | | | | | |
| 5 | | | | | |
| 6 | | | | | |
| 7 | | Heterozygous | | | |
| 8 | | | | | |
| 9 | | | | | |
| 10 | | | | | |
| 11 | | | | | |
| 12 | | | | | |
| 13 | | | | | |
| 14 | | | | | |
| 15 | | | | | |
| 16 | | | | | |

## Concept Mapping

The construction of and theory behind concept mapping are discussed on pages vii–ix in the front of this Study Guide. Read those pages carefully. Then consider the concepts presented in Section 9–1 and how you would organize them into a concept map. Now look at the concept map for Chapter 9 on page 93. Notice that the concept map has been started for you. Add the key facts and concepts you feel are important for Section 9–1. When you have finished the chapter, you will have a completed concept map.

## SECTION REVIEW

In the first part of this section you learned that Mendel's work was not appreciated during his lifetime. The probable reason for this is that Mendel had devised a totally new and different approach to studying biology—an approach that other biologists did not understand.

One of the most innovative things that Mendel did was to apply probability to biology. Recall that probability is the likelihood that a particular event will occur. The probability of an event, such as a particular outcome of a cross, can be determined mathematically.

However, you can calculate only what is *likely* to happen. You cannot tell exactly what the outcome of a single event will be. Two basic rules of probability are (1) You get the expected results only for large numbers of trials and (2) previous events do not affect future outcomes.

In the second part of the chapter you learned that a Punnett square can be used to analyze the probable results of a cross. You also learned how to use Punnett squares to solve genetics problems.

### Defining Terms: Building Vocabulary Skills

In your own words, define each of the following terms.

**1.** Probability: _____

_____

_____

_____

_____

_____

_____

_____

_____

**2.** Test cross: _____

_____

_____

_____

_____

_____

_____

_____

### Using Punnett Squares: Practicing Skills

| SELECTED TRAITS IN CATS | | |
|---|---|---|
| **Trait** | **Dominant Allele** | **Recessive Allele** |
| **Coat length** | Short hair (*H*) | Long hair (*h*) |
| **Tabby stripes** | Tabby (*T*) | Stripeless (*t*) |
| **Colorpoint (markings on nose, ears, paws, and tail)** | Normal (no colorpoint) (*N*) | Colorpoint (*n*) |

Show the results of the following crosses using Punnett squares and the information in the accompanying figure.

1. Heterozygous short-hair X
   heterozygous short-hair


   Genotypic ratio:
   Phenotypic ratio:

2. Heterozygous tabby X stripeless


   Genotypic ratio:
   Phenotypic ratio:

3. Colorpoint X homozygous normal


   Genotypic ratio:
   Phenotypic ratio:

**4.** Homozygous short, homozygous colorpoint X
homozygous long, homozygous normal

Genotypic ratio:
Phenotypic ratio:

**5.** Heterozygous short, heterozygous normal X
heterozygous short, heterozygous normal

Genotypic ratio:
Phenotypic ratio:

**6.** Heterozygous tabby, heterozygous normal X
stripeless colorpoint

Genotypic ratio:
Phenotypic ratio:

**7.** Long-hair, heterozygous normal X
Long-hair, heterozygous normal

Genotypic ratio:
Phenotypic ratio:

## Concept Mapping

The construction of and theory behind concept mapping are discussed on
pages vii–ix in the front of this Study Guide. Read those pages carefully. Then
consider the concepts presented in Section 9–2 and how you would organize
them into a concept map. Now look at the concept map for Chapter 9 on
page 93. Notice that the concept map has been started for you. Add the key
facts and concepts you feel are important for Section 9–2. When you have
finished the chapter, you will have a completed concept map.

**Section 9–3**  **Meiosis**  *(pages 193–196)*

### SECTION REVIEW

In this section you discovered why gametes are not formed by mitosis. You also learned about the process of meiosis. Meiosis is a special type of cell division that produces gametes. In most organisms meiosis takes place in two stages. In the first stage of meiosis homologous chromosomes pair up and may exchange segments in a process called crossing-over. Then the homologous pairs line up along the equator of the cell. Finally, the homologous chromosomes separate and two new cells are formed. These two cells have half as many chromosomes as the parent cell and thus only half the genetic information. In the second stage of meiosis the chromosomes line up in the center of each cell. The centromere divides, the chromatids are separated, and each cell divides to form two new cells. Note that meiosis produces four haploid cells that are different from one another and from the original parent cell.

Later in this section you related the events of meiosis to the principles of genetics. Recall that the separation of homologous chromosomes during Meiosis I results in segregation. And the random shuffling of chromosomes to different cells results in independent assortment. Next, you learned how gamete formation differs in males and females. Finally, you compared meiosis and mitosis.

**Matching Terms: Building Vocabulary Skills**

In the space provided, write the term that best fits the description.

_____  **1.** Contains both sets of homologous chromosomes

_____  **2.** Cell division that produces gametes

_____  **3.** N

_____  **4.** 2 chromosomes that carry alleles for the same traits

_____  **5.** Contains a single set of chromosomes

_____  **6.** Process in which homologs exchange portions of their chromatids

_____  **7.** Describes corresponding chromosomes

_____  **8.** Small cell resulting from meiosis in female animals, which usually does not participate in reproduction

_____  **9.** Male gamete in higher plants

### Reviewing Concepts: Finding the Main Ideas

**1.** In your own words, compare mitosis and meiosis. _____

_____

_____

_____

_____

_____

**2.** How is meiosis related to gamete formation? _____

_____

_____

**3.** In which stage of meiosis does crossing-over occur? Why is crossing-over important?

_____

_____

_____

**4.** How does gamete formation differ in male and female animals?

_____

_____

_____

_____

**5.** How does meiosis result in segregation of alleles? _____

_____

_____

_____

_____

### Concept Mapping

The construction of and theory behind concept mapping are discussed on pages vii–ix in the front of this Study Guide. Read those pages carefully. Then consider the concepts presented in Section 9–3 and how you would organize them into a concept map. Now look at the concept map for Chapter 9 on page 93. Notice that the concept map has been started for you. Add the key facts and concepts you feel are important for Section 9–3. When you have finished the chapter, you will have a completed concept map.

## Using the Writing Process                                           *Chapter 9*

Use your writing skills and imagination to respond to the following writing assignment. You will probably need an additional sheet of paper to complete your response.

Imagine that you are the *t* allele in a short pea plant. Describe a *tt* X *TT* cross and a subsequent *Tt* X *Tt* cross from your point of view. In your description be sure totell what happens during meiosis to the chromosome on which you are located.

*Hint:* Here are some questions you might wish to answer in your description. How does it feel to be relocated by crossing-over? Are you sad to be separated from alleles on other chromosomes as a result of segregation and independent assortment, or are you glad to be rid of them? Do you like the alleles that you meet as a result of fertilization? What do you think of your corresponding dominant allele?

_____

_____

_____

_____

_____

_____

_____

_____

_____

_____

_____

_____

_____

_____

_____

_____

_____

_____

## Concept Mapping

The concept map below has been started for you. Add the key facts and concepts for each section of the chapter to its particular concept map. When you are done, you will have a concept map for the entire chapter.

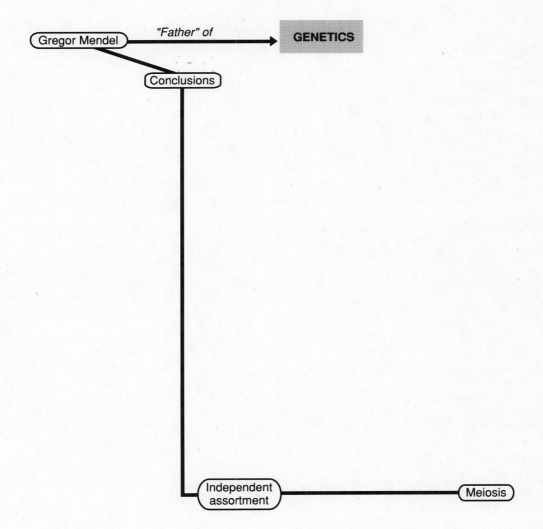

S T U D Y
G U I D E

CHAPTER
*Genes and Chromosomes* **10**

| Section 10–1 | **The Chromosome Theory of Heredity** | *(pages 205–211)* |

## SECTION REVIEW

At the beginning of this section you read about the evidence that led Sutton to formulate the chromosome theory of heredity in 1902. This theory states that genes, the factors that control heredity, are located on the chromosomes. Each gene occupies a specific place on a chromosome, and each chromosome contains one allele for each gene.

Next, you were introduced to Morgan's experiments with the fruit fly *Drosophila melanogaster*, and you learned about gene linkage, crossing-over, and gene mapping. Genes that are located on the same chromosome are said to be linked. Linked genes do not undergo independent assortment. Gene

linkages can be broken by the process of crossing-over, which occurs during meiosis. The farther apart the genes are located on a chromosome, the greater the frequency of crossing-over between them. Recall that you can map the position of genes on a chromosome if you know the frequency of crossing-over between pairs of genes.

Finally, you learned about sex chromosomes and discovered how sex-linked traits are inherited. Recall that sex chromosomes are paired chromosomes that are different in males and in females. Genes located on the sex chromosomes are said to be sex-linked.

### Completing Sentences: Building Vocabulary Skills

In the space provided, write the term that best completes the statement.

**1.** Genes that are located on the sex chromosomes are said to be _____.

**2.** Crossing-over produces _____, or individuals with new combinations of genes.

**3.** The _____ states that genes are located on chromosomes.

**4.** _____, "packages" of genes that always tend to be inherited together, correspond to chromosomes.

**5.** The fact that the distance between two genes determines how often crossing-over occurs between them is the basis for _____.

**6.** _____ are chromosomes that are different in males and in females.

**7.** Chromosomes that are the same in both males and females are known as _____.

**8.** _____, which are located on the same chromosome, do not undergo independent assortment.

**9.** If we consider just the sex chromosomes, a normal female human or fruit fly is represented as _____, whereas a normal male is represented as _____.

**10.** In humans and fruit flies, the _____ parent is responsible for determining the sex of the offspring.

**11.** The _____ in fruit flies is small and hook-shaped and does not resemble its corresponding chromosome.

**12.** In humans and fruit flies, the effects of sex-linked recessive alleles tend to show up more

frequently in _____.

### Gene Mapping: Applying the Main Ideas

The following table gives the frequency of crossing-over between 7 genes on the X chromosome in *Drosophila*. Use the information in the table to draw a gene map for the X chromosome.

| Genes | Frequency of Crossing-over |
|-------|---------------------------|
| Yellow body; vermilion eyes | 33 % |
| Bobbed hairs; vermilion eyes | 33 % |
| White eyes; garnet eyes | 42.5% |
| Yellow body; white eyes | 1.5% |
| Vermilion eyes; garnet eyes | 11 % |
| Bobbed hairs; singed bristles | 45 % |
| Ruby eyes; singed bristles | 13.5% |
| Ruby eyes; garnet eyes | 36.5% |
| Bobbed hairs; garnet eyes | 22 % |
| White eyes; ruby eyes | 6 % |

| Gene Map for the X Chromosome in *Drosophila* |
|---|
| |

### Sex Linkage: Applying the Main Ideas

1. In fruit flies, red eyes ($X^R$) are dominant over white eyes ($X^r$). In the space provided, draw a Punnett square that shows the results of a cross between a white-eyed male and a heterozygous red-eyed female.

Phenotypic ratio: _____

Genotypic ratio: _____

In fruit flies and humans and other mammals, sex is determined by an X-Y system. However, many organisms do not have the X-Y system of sex determination. For example, birds have a Z-W system. Male birds are ZZ, whereas female birds are ZW.

2. In birds, which parent determines the sex of its offspring? Why?

_____

_____

3. In chickens, barred feathers ($Z^B$) are dominant over nonbarred feathers ($Z^b$).

   a. Draw a Punnett square that shows the results of a cross between a barred female and a nonbarred male.

   % Barred females: _____

   % Nonbarred females: _____

   % Barred males: _____

   % Nonbarred males: _____

   b. Draw a Punnett square that shows the results of a cross between a nonbarred female and a heterozygous barred male.

   % Barred females: _____

   % Nonbarred females: _____

   % Barred males: _____

   % Nonbarred males: _____

   ratio: _____

### Concept Mapping

The construction of and theory behind concept mapping are discussed on pages vii–ix in the front of this Study Guide. Read those pages carefully. Then consider the concepts presented in Section 10–1 and how you would organize them into a concept map. Now look at the concept map for Chapter 10 on page 104. Notice that the concept map has been started for you. Add the key facts and concepts you feel are important for Section 10–1. When you have finished the chapter, you will have a completed concept map.

**SECTION REVIEW**

In this section you learned about mutations. A mutation is a change in the genetic information of a cell. Many mutations either have no effect or cause slight, harmless changes. A few mutations are harmful. And once in a while a mutation may be beneficial.

Mutations that affect the reproductive cells are known as germ mutations. They can be inherited. Mutations that affect nonreproductive cells are known as somatic mutations. They cannot be inherited. Both germ and somatic mutations may occur at either the level of chromosomes or the level of genes.

Mutations that involve segments of chromosomes, whole chromosomes, or entire sets of chromosomes are called chromosomal mutations. Chromosome structure is altered or rearranged by deletion, duplication, inversion, and translocation mutations. Chromosome number is changed by nondisjunction.

Mutations that involve individual genes are called gene mutations. Some gene mutations involve many nucleotides. Others, known as point mutations, affect only one nucleotide. A point mutation that involves the substitution of one base for another may have little or no effect. But one that involves the addition or deletion of a base can completely change the polypeptide produced by a gene.

**Comparing Terms: Building Vocabulary Skills**

Explain how the paired terms differ from each other.

**1.** Chromosomal mutation, gene mutation: _____

_____

_____

_____

_____

_____

_____

_____

**2.** Somatic mutation, germ mutation: _____

_____

_____

_____

_____

_____

_____

**3.** Nondisjunction, polyploidy: _____

_____

_____

_____

_____

_____

**4.** Crossing-over, translocation: _____

_____

_____

_____

_____

_____

**5.** Point mutation, inversion: _____

_____

_____

_____

_____

_____

**6.** Frameshift mutation, duplication mutation: _____

_____

_____

_____

_____

_____

### Concept Mapping

The construction of and theory behind concept mapping are discussed on pages vii–ix in the front of this Study Guide. Read those pages carefully. Then consider the concepts presented in Section 10–2 and how you would organize them into a concept map. Now look at the concept map for Chapter 10 on page 104. Notice that the concept map has been started for you. Add the key facts and concepts you feel are important for Section 10–2. When you have finished the chapter, you will have a completed concept map.

**Section 10–3** | **Regulation of Gene Expression** | *(pages 215–221)*

## SECTION REVIEW

In this section you were introduced to the ways in which the expression and activity of genes are regulated and controlled. Keep in mind that individual genes do not function in isolation. Instead, they interact with other genes and with their environment.

First, you were introduced to gene interaction. You also learned about the chemical basis for these interactions. Recall that the terms *dominance, incomplete dominance,* and *codominance* describe ways in which the two alleles for a particular gene interact. The term *polygenic inheritance* refers to traits that are controlled by two or more genes.

Next, you learned about one mechanism in prokaryotes by which the genes in an operon are turned on and off. Recall that an operon consists of a cluster of genes and two

regions that control these genes: the operator and the promoter. The genes in the operon that you examined in this section produce enzymes that break down the sugar lactose. When lactose is present, the genes are expressed. When lactose is not present, a molecule called the repressor attaches to the operator region of the operon and turns the genes off.

Finally, you read about some of the ways in which genes are regulated in eukaryotes. One way of controlling which genes are expressed involves the need to process RNA before it can leave the nucleus as mRNA. Such processing involves removing the introns (intervening sequences) from pre-mRNA and splicing together the remaining exons (expressed sequences).

### Writing Paragraphs: Building Vocabulary Skills

Using your own words, write 1 or 2 paragraphs comparing gene regulation in prokaryotes and in eukaryotes on a separate sheet of paper. Use each of the terms in the following list at least once. You may use plural forms of nouns and different tenses of verbs. Underline each term every time it is used.

| Terms That Should Appear in the Paragraph(s) | | |
|---|---|---|
| DNA | intron | pre-mRNA |
| eukaryote | mRNA | prokaryote |
| exon | operator | promoter |
| gene | operon | repressor |
| inducer | polypeptide | RNA polymerase |

### Gene Interactions: Applying the Main Ideas

1. In horses, the dominant *D* allele allows coat color to develop fully. The recessive *d* allele dilutes, or weakens, coat color. When a chestnut (brown) horse is crossed with a cremello (nearly white) horse, their offspring are palominos (gold with white manes and tails.)

   **a.** What kind of gene interaction is illustrated by the *D* gene? Explain.

   _____

   _____

   _____

   **b.** What is the genotype of a chestnut horse? _____

   A cremello horse? _____

   A palomino horse? _____

   **c.** Fill in the following Punnett square to show the results of a cross between two palomino horses.

   % chestnut:_____

   % palomino: _____

   % cremello:_____

   genotypic ratio: _____

2. Some plants, such as cherries and petunias, have a gene called *S* that prevents self-pollination. If the *S* allele in a pollen grain matches either of the two *S* alleles in a flower, then it cannot fertilize the egg cells in the flower. For example, an $S_1$ pollen grain cannot fertilize the eggs in an $S_1S_2$ flower. But it can fertilize the eggs in an $S_3S_4$ flower.

   **a.** What kind of gene interaction is illustrated by the *S* gene? Explain.

   _____

   _____

   _____

   _____

   _____

   **b.** Why do egg cells and pollen cells have only one *S* allele?

   _____

   _____

   _____

   _____

   _____

**c.** The *S* gene is an example of multiple alleles. Why is the term multiple alleles a good

one?_____

_____

_____

_____

**3.** Mice, horses, and many other mammals may be black (*B*), brown (*b*), spotted (*T*), or unspotted (*t*).

**a.** What kind of gene interaction is illustrated by the *B* and *T* genes? Explain.

_____

_____

_____

_____

**b.** Fill in the following Punnett square to show the cross *BbTt* X *Bbtt*. Then circle each genotype that produces an unspotted brown individual.

|  |  |  |  |
|---|---|---|---|
|  |  |  |  |
|  |  |  |  |
|  |  |  |  |
|  |  |  |  |

**c.** Why was the letter *T*, rather than *S*, chosen to represent the gene for spots?

_____

_____

_____

_____

### Concept Mapping

The construction of and theory behind concept mapping are discussed on pages vii–ix in the front of this Study Guide. Read those pages carefully. Then consider the concepts presented in Section 10–3 and how you would organize them into a concept map. Now look at the concept map for Chapter 10 on page 104. Notice that the concept map has been started for you. Add the key facts and concepts you feel are important for Section 10–3. When you have finished the chapter, you will have a completed concept map.

## Using the Writing Process

Use your writing skills and imagination to respond to the following writing assignment. You will probably need an additional sheet of paper to complete your response.

Imagine that you are a young enzyme seeking your fortune in a bacterial cell far from home. You have just been hired to work as the repressor for an important operon. Write a letter to the folks back home that explains your new job and describes your first day at work. Discuss any hopes and concerns you might have regarding your job as a repressor.

_____

_____

_____

_____

_____

_____

_____

_____

_____

_____

_____

_____

_____

_____

_____

_____

_____

_____

## Concept Mapping

The concept map below has been started for you. Add the key facts and concepts for each section of the chapter to this partial concept map. When you are done, you will have a concept map for the entire chapter.

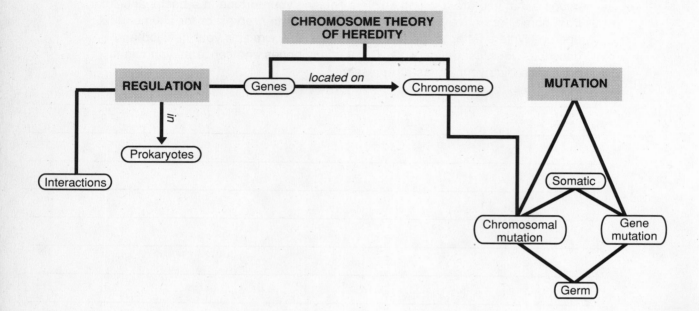

S T U D Y
G U I D E

| Section 11-1 | "It Runs in the Family" | (pages 227-229) |

### SECTION REVIEW

This section presented some basic facts and concepts that you need in order to understand human heredity. You learned that many human traits are determined by the same kind of gene interactions that occur in pea plants, fruit flies, and other organisms. And, as in other organisms, the expression of certain genes is strongly influenced by environmental factors such as nutrition.

A human diploid cell contains 46 chromosomes arranged in 23 pairs: 22 pairs of autosomes and one pair of sex chromosomes. A human gamete, which is haploid, contains 23 chromosomes. The diploid number of chromosomes is restored when a sperm cell and an egg cell join during fertilization to form a zygote.

### Understanding Concepts: Finding the Main Ideas

1. How do gametes differ from other cells in the human body?

_____

_____

2. What happens to the chromosome number during fertilization in humans?

_____

_____

_____

3. Explain the following observation: "Human heredity is a combination of nature and

nurture." _____

_____

_____

### Concept Mapping

The construction of and theory behind concept mapping are discussed on pages vii–ix in the front of this Study Guide. Read those pages carefully. Then consider the concepts presented in Section 11–1 and how you would organize them into a concept map. Now look at the concept map for Chapter 11 on page 115. Notice that the concept map has been started for you. Add the key facts and concepts you feel are important for Section 11–1. When you have finished the chapter, you will have a completed concept map.

## SECTION REVIEW

In this section, you examined the inheritance of several specific human traits. First, you studied how ABO and Rh blood groups are inherited. ABO blood groups, which are of particular importance in blood transfusions, are determined by multiple alleles. Two alleles, $I^A$ and $I^B$, are codominant. One allele, $i$, is recessive. Rh blood groups are determined by a dominant Rh positive allele and a recessive Rh negative allele. Next, you learned about Huntington disease, which is caused by a dominant allele. You then learned about sickle cell anemia, which is caused by codominant alleles: one for normal hemoglobin and one for sickle cell hemoglobin. Sickle cell hemoglobin crystallizes when oxygen is in short supply, causing red blood cells to become sickle-shaped and rigid. The sickle-shaped blood cells tend to become stuck in capillaries, blocking the flow of blood and thus damaging cells and tissues. Finally, you read about polygenic traits in humans. Polygenic traits include height and skin color.

### ABO Blood Groups: Using the Main Ideas

Use the space provided to the side of the following genetics problems to draw Punnett squares to help you solve the problems.

**1.** A man with type O blood and a woman with type AB blood get married.

  **a.** What is the probability that they will have a child with type A blood?_____

  **b.** Suppose that one of the couple's children needs an operation. This child has type B blood. Can the child safely receive a blood transfusion from either parent? Explain. _____

    _____

    _____

    _____

    _____

    _____

**2.** Vincent has type A blood and his mother has type O blood. Christine has type B blood and her father has type O blood.

  **a.** What is Vincent's genotype? _____

  **b.** What is Christine's genotype? _____

**c.** What genotype(s) might Christine's mother

have? _____

_____

_____

**d.** Suppose Vincent and Christine get married. What

is the probable phenotypic ratio for their

offspring? _____

_____

_____

**e.** What is the probable genotypic ratio for their

offspring? _____

_____

### Genetic Disorders: Applying the Main Ideas

In the space provided to the side of each of the following genetics
problems, draw a Punnett square to help you solve the problem. Then
answer the questions.

1. The allele for normal hemoglobin can be
represented as $H^A$. The allele for the sickle cell
hemoglobin can be represented as $H^S$.

**a.** What type of gene interaction is involved in

sickle cell anemia? _____

_____

Consider the offspring of two people who both
have the genotype $H^A H^S$.

**b.** What percentage of their offspring are likely to

be sickle cell sufferers? _____

**c.** What percentage of their offspring are likely to

be resistant to malaria and suffer few effects of

the disease? _____

2. **a.** How is Huntington disease inherited?

_____

_____

**b.** What is the probability that an individual who has one parent with Huntington disease will also have the disease? (Assume the other parent does

not have the disease.) _____

3. Phenylketonuria (PKU) is a genetic disease in which the body cannot safely break down the amino acid phenylalanine. If untreated, PKU causes severe brain damage. To avoid this, people with PKU must eat a special diet low in phenylalanine.

   Two people who have normal phenotypes have a child. A blood test at birth shows that the child has PKU.

   **a.** How is PKU inherited? Explain. _____

_____

_____

_____

   **b.** What is the probability that this couple's next

child will have PKU? _____

   **c.** What is the probability that this couple's next child will be homozygous for the normal allele?

_____

4. Achondroplasia, a form of dwarfism, is caused by the dominant allele *A*. Embryos with the genotype *AA* do not survive. Suppose that two people with achondroplasia get married and have children.

   **a.** What phenotypic ratio would you probably observe in the couple's children?

_____

_____

_____

_____

_____

_____

**b.** One of this couple's children, who has a normal phenotype, marries a person who also has a normal phenotype. What percentage of the children from this marriage are likely to have

achondroplasia? Explain. _____

_____

_____

_____

_____

_____

5. The disease cystic fibrosis is caused by the recessive allele *n*.

**a.** What percentage of the children of a couple with genotypes *NN* and *Nn* will probably have cystic fibrosis? Explain.

_____

_____

_____

_____

**b.** What are the phenotypic and genotypic ratios for the offspring of two people who both have the genotype *Nn?*

_____

_____

_____

_____

_____

## Concept Mapping

The construction of and theory behind concept mapping are discussed on pages vii–ix in the front of this Study Guide. Read those pages carefully. Then consider the concepts presented in Section 11–2 and how you would organize them into a concept map. Now look at the concept map for Chapter 11 on page 115. Notice that the concept map has been started for you. Add the key facts and concepts you feel are important for Section 11–2. When you have finished the chapter, you will have a completed concept map.

**Sex-Linked Inheritance** *(pages 235–239)*

## SECTION REVIEW

In this section you learned about patterns of human inheritance associated with the sex chromosomes. These are sex determination, sex-linked disorders, and sex-influenced traits.

In humans, the sex of an individual is determined by whether an egg cell, which contains an X chromosome, is fertilized by an X-carrying sperm or a Y-carrying sperm. If just the sex chromosomes are considered, a female is normally XX and a male is normally XY. By studying nondisjunction disorders that involve the sex chromosomes, it has been learned that at least one X chromosome is necessary for survival and that maleness in humans is determined by the presence of a Y chromosome. (Recall that in fruit flies sex is determined by the number of X chromosomes.)

Genes that are carried on the sex chromosomes are said to be sex-linked. Defective recessive alleles located on the X chromosome are expressed more commonly in males than in females. Examples of genetic disorders caused by recessive genes on the X chromosome include colorblindness, hemophilia, and Duchenne muscular dystrophy.

Sex-influenced traits, such as male pattern baldness, are caused by genes located on autosomes. However, these genes are expressed differently in males and females.

## Defining Terms: Building Vocabulary Skills

In your own words, define each of the following. Use complete sentences.

1. Nondisjunction: _____

_____

_____

2. Sex-linked trait: _____

_____

_____

3. Sex-influenced trait: _____

_____

_____

4. Turner syndrome: _____

_____

_____

5. Colorblindness: _____

_____

_____

### Understanding Sex-Linked Inheritance: Applying the Main Ideas

1. Duchenne muscular dystrophy is caused by a recessive allele located on the X chromosome.

   **a.** In the space provided, draw a Punnett square that shows how two unaffected people can have a child who has Duchenne muscular dystrophy.

   Genotypic ratio: _____

   _____

   Phenotypic ratio:_____

   _____

   **b.** Is the child with muscular dystrophy male or female? _____

   **c.** What is the probability that the couple's next child will have muscular dystrophy?

   _____

   **d.** What percentage of the couple's male children will probably have muscular

   dystrophy? _____

   **e.** What percentage of the couple's female children will probably have muscular

   dystrophy?_____

2. Hemophilia is caused by a recessive allele on the X chromosome.

   **a.** In the space provided, draw a Punnett square that shows how a mother whose blood clots normally can have a daughter with hemophilia.

   Genotypic ratio:_____

   _____

   _____

   Phenotypic ratio: _____

   _____

   _____

   **b.** What is the father's genotype and phenotype? _____

   _____

   **c.** Why is it extremely uncommon for a female to have hemophilia?

   _____

   _____

   _____

   _____

**3.** Not all sex-linked genes are recessive. For example, hypophosphatemia is caused by a dominant allele on the X chromosome.

**a.** Using *A* to represent the allele for hypophosphatemia, draw a Punnett square that shows the possible offspring of a woman with genotype $X^A X^a$ and a man with genotype $X^a Y$.

**b.** What percentage of the children have hypophosphatemia? _____

**c.** Are males more likely to have hypophosphatemia than females? Explain.

_____

_____

_____

**d.** Draw a Punnett square that shows the possible offspring of a woman with a normal phenotype and a man with hypophosphatemia.

Genotypic ratio: _____

_____

Phenotypic ratio: _____

_____

_____

**e.** What percentage of the children have hypophosphatemia? _____

**f.** What percentage of the male offspring have hypophosphatemia? Explain this result.

_____

_____

_____

_____

**Concept Mapping**

The construction of and theory behind concept mapping are discussed on pages vii–ix in the front of this Study Guide. Read those pages carefully. Then consider the concepts presented in Section 11–3 and how you would organize them into a concept map. Now look at the concept map for Chapter 11 on page 115. Notice that the concept map has been started for you. Add the key facts and concepts you feel are important for Section 11–3. When you have finished the chapter, you will have a completed concept map.

## Section 11–4    Diagnosis of Genetic Disorders                    *(pages 239–241)*

### SECTION REVIEW

Genetic disorders may result from chromosomal abnormalities that affect the autosomes as well as those that affect sex chromosomes. In many cases these disorders may be detected by examining under a microscope the cells from a developing embryo. The cells may come from the fluid surrounding the embryo, as in the case of amniocentesis, or they may come from the sac that contains the fluid, as in chorionic villus biopsy. Other genetic disorders may be detected by tests for biochemical abnormalities or for the presence of certain DNA sequences.

You may know a person who is affected by Down syndrome, which usually results from nondisjunction of chromosome 21. Affected individuals have an extra copy of chromosome 21 or an extra portion of this chromosome. You are probably aware that Down syndrome results in mental retardation, which may range from mild to severe.

Our ability to detect genetic disorders before birth has led to serious moral and ethical questions. As adult members of our society, you will no doubt be called upon to deal with these issues.

### Diagnosing Genetic Disorders: Applying the Main Ideas

1. What is a karyotype?_____

_____

_____

2. What causes Down syndrome? _____

_____

_____

3. How are amniocentesis and chorionic villus biopsy used in diagnosing prenatal

disorders? _____

_____

_____

_____

### Concept Mapping

The construction of and theory behind concept mapping are discussed on pages vii–ix in the front of this Study Guide. Read those pages carefully. Then consider the concepts presented in Section 11–4 and how you would organize them into a concept map. Now look at the concept map for Chapter 11 on page 115. Notice that the concept map has been started for you. Add the key facts and concepts you feel are important for Section 11–4. When you have finished the chapter, you will have a completed concept map.

## Using the Writing Process                    *Chapter 11*

Use your writing skills and imagination to respond to the following writing assignment. You will probably need an additional sheet of paper to complete your response.

Write a short story or a dramatic monologue in which a human X chromosome reminisces about its existence.

*Hint:* How did the X chromosome affect the lives of parents and offspring? What does it think about other chromosomes? How was the X chromosome involved in sex determination? Was it involved in any genetic disorders?

_____

_____

_____

_____

_____

_____

_____

_____

_____

_____

_____

_____

_____

_____

_____

_____

_____

_____

_____

_____

## Concept Mapping

The concept map below has been started for you. Add the key facts and concepts for each section of the chapter to this partial concept map. When you are done, you will have a concept map for the entire chapter.

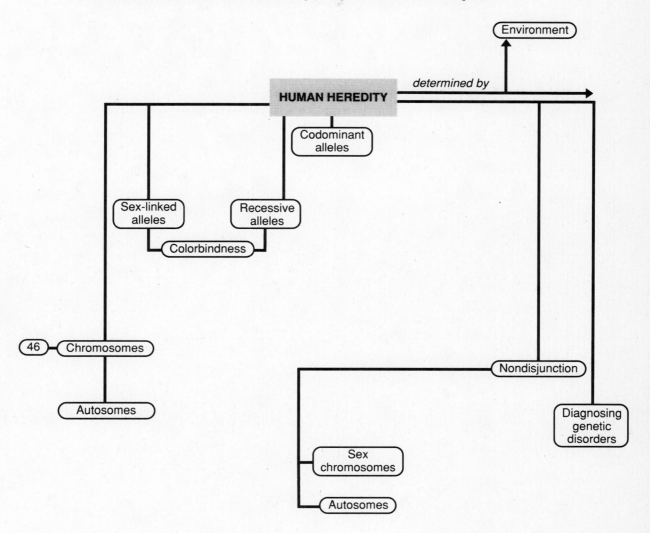

S T U D Y
G U I D E

CHAPTER
*Genetic Engineering* **12**

---

| Section 12–1 | **Modifying the Living World** | *(pages 247–250)* |

### SECTION REVIEW

It should not be surprising to you that people have tried to improve the world around them by attempting to make "better" living things. Even before their genetic basis was understood, breeding techniques were used to produce varieties of plants and animals that had desirable traits. In selective breeding, the individuals that have the "best" traits are chosen to produce the next generation. In inbreeding, individuals with similar characteristics are crossed so that these characteristics will appear in their offspring. In hybridization, dissimilar individuals are crossed in the hope that the desired characteristics of each parent will combine in the offspring.

Recall that selective breeding is confined to characteristics already in the population. New characteristics are produced by mutations. Most mutations have little or no effect, and some mutations are harmful. But because a few mutations are beneficial, a breeder may try to cause mutations. If the breeder is lucky and obtains a beneficial mutation, selective breeding can be used to obtain an entire population that has the new characteristic.

### Applying Definitions: Building Vocabulary Skills

In the space provided, define each of the breeding techniques in your own words. Then give a specific example of an organism that has been developed using the technique.

**1.** Inbreeding: _____

_____

_____

_____

**2.** Selective breeding: _____

_____

_____

_____

**3.** Mutagenesis: _____

_____

_____

_____

### ▨ Relating Cause and Effect: Using the Main Ideas

Selective breeding, inbreeding, hybridization, and mutagenesis have been used to produce many varieties of organisms. In the space provided, write the name of the technique that is most closely associated with each of the following descriptions.

**1.** Navel (seedless) orange trees were originally found in Brazil over 100 years ago.

_____

**2.** Mules, which are produced by crossing horses and donkeys, are stronger than donkeys and are hardier than horses. _____

**3.** Breeders of miniature horses usually choose the smallest horses as parents of the next generation. _____

**4.** The high-protein grain triticale was produced by crossing wheat and rye plants.

_____

**5.** Useful bacterial strains have been produced by treating bacteria with chemicals or radiation. _____

**6.** The first Himalayan cat was produced from a cross between a female cat with long black fur and her male parent. _____

### ▨ Picking Puppies: Using the Main Ideas

Purebred puppies are usually placed into three categories: show, breeder, and pet. Show-quality puppies exhibit a combination of good qualities that make it likely that they will be dog show winners when they are old enough to compete. Breeder-quality puppies are those that have good qualities but are not likely to win trophies and ribbons. Pet-quality puppies possess characteristics that are considered undesirable in their breed. For example, the color of their coat may be "wrong." Pet-quality puppies are usually much less expensive than show or breeder puppies.

**1.** Why are show-quality puppies often much more expensive than other puppies in the same litter? _____

_____

_____

_____

_____

_____

_____

**2.** Some breeders will sell a pet-quality puppy without its "papers"—its pedigree and other certificates. Other breeds insist that a pet-quality dog be sterilized, or made unable to have offspring, before they give the new owner its papers. Explain these practices.

_____

_____

_____

_____

_____

_____

**3.** Why is it generally considered a bad idea to buy a purebred puppy if its parents lack

papers? _____

_____

_____

_____

_____

_____

_____

**4.** Explain why a person might prefer to buy a pet-quality puppy. _____

_____

_____

_____

_____

_____

_____

### Concept Mapping

The construction of and theory behind concept mapping are discussed on pages vii–ix in the front of this Study Guide. Read those pages carefully. Then consider the concepts presented in Section 12–1 and how you would organize them into a concept map. Now look at the concept map for Chapter 12 on page 124. Notice that the concept map has been started for you. Add the key facts and concepts you feel are important for Section 12–1. When you have finished the chapter, you will have a completed concept map.

**Section 12–2**  **Genetic Engineering: Technology and Heredity**  *(pages 251–256)*

### SECTION REVIEW

In this section you were introduced to genetic engineering. Genetic engineering is a set of techniques that allow researchers to directly manipulate DNA.

In the first part of this section you learned about the basic techniques that are used in genetic engineering. Recall that restriction enzymes are used to cut DNA at specific nucleotide sequences. The resulting DNA fragments can then be recombined with an organism's genetic material and inserted into living cells. Many copies of a particular DNA fragment can be produced by DNA cloning. The sequence of nucleotides in a cloned DNA fragment is determined by a process that involves chemical treatments and electrophoresis.

In the second part of this section you read about some specific examples of genetic engineering in bacteria, in plants, and in animals. Organisms that contain foreign genes are said to be transgenic. They are also known as chimeras.

### Relating Terms: Building Vocabulary Skills

In your own words, define the terms in each of the following pairs. Then explain how the paired terms are related to each other.

**1.** Restriction enzyme, plasmid: _____

_____

_____

_____

**2.** Recombinant DNA, chimera: _____

_____

_____

_____

**3.** DNA insertion, clone: _____

_____

_____

_____

**4.** DNA sequencing, electrophoresis: _____

_____

_____

_____

**5.** Genetic engineering, transgenic: _____

_____

_____

_____

### Explaining Concepts: Using the Main Ideas

**1.** The protein AHF (**anti**hemophilic **f**actor) is needed for blood to clot normally. In the genetic disorder hemophilia, AHF is missing. Describe how bacteria could be used to produce a plentiful supply of AHF for people with hemophilia.

_____

_____

_____

_____

_____

_____

_____

**2.** In nature, the protein luciferin is found only in animals that can produce light, such as fireflies. In the presence of ATP and the enzyme luciferase, luciferin combines with oxygen and produces a bright greenish light.

**a.** How might you test living aerobic cells for the presence of luciferase?

_____

_____

**b.** Explain why molecular biologists might splice the gene for luciferase onto the gene

they wish to study. _____

_____

_____

### Concept Mapping

The construction of and theory behind concept mapping are discussed on pages vii–ix in the front of this Study Guide. Read those pages carefully. Then consider the concepts presented in Section 12–2 and how you would organize them into a concept map. Now look at the concept map for Chapter 12 on page 124. Notice that the concept map has been started for you. Add the key facts and concepts you feel are important for Section 12–2. When you have finished the chapter, you will have a completed concept map.

| Section 12–3 | The New Human Genetics | (pages 256–259) |

## SECTION REVIEW

In this section you discovered that molecular biology has a number of interesting applications in humans. Among these are sequencing of the human genome, finding the genes for inherited disorders, and identifying criminals through DNA fingerprinting. One potential application of genetic engineering includes curing genetic diseases by replacing defective genes. Although these applications promise to improve human health and expand our understanding of human genetics, their use raises many ethical issues. It is clear that society as a whole must be prepared to deal with these issues in an intelligent, informed, and humane manner.

### Defining Terms: Building Vocabulary Skills

In your own words, define each of the following terms.

**1.** Genome: _____

_____

_____

_____

**2.** DNA fingerprinting: _____

_____

_____

_____

_____

**3.** Gene therapy: _____

_____

_____

_____

### Concept Mapping

The construction of and theory behind concept mapping are discussed on pages vii–ix in the front of this Study Guide. Read those pages carefully. Then consider the concepts presented in Section 12–3 and how you would organize them into a concept map. Now look at the concept map for Chapter 12 on page 124. Notice that the concept map has been started for you. Add the key facts and concepts you feel are important for Section 12–3. When you have finished the chapter, you will have a completed concept map.

## Using the Writing Process                                    *Chapter 12*

Use your writing skills and imagination to respond to the following writing assignment. You will probably need an additional sheet of paper to complete your response.

> Imagine that you are a member of the United States Congress. You have been appointed to head the committee that will draft a bill proposing laws for governing genetic engineering. To prepare for drafting this bill and participating in the associated debates, make a list of the laws that you wish to propose. Note why you think each of your laws should be put into effect.

_____

_____

_____

_____

_____

_____

_____

_____

_____

_____

_____

_____

_____

_____

_____

_____

_____

_____

_____

_____

_____

_____

_____

_____

## Concept Mapping *Chapter 12*

The concept map below has been started for you. Add the key facts and concepts for each section of the chapter to this partial concept map. When you are done, you will have a concept map for the entire chapter.

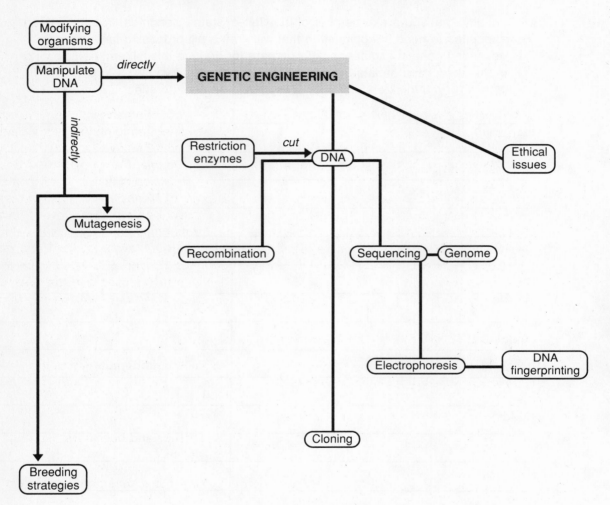

S T U D Y
G U I D E

CHAPTER **13**
*Evolution: Evidence of Change*

| Section 13-1 | **Evolution and Life's Diversity** | *(pages 269-271)* |

### SECTION REVIEW

In this section you were introduced to the concept of evolution. Evolution is the process by which modern organisms have descended from ancient organisms. In other words, evolution is change over time.

The person who contributed the most to our understanding of the process of evolution is Charles Darwin. When Darwin was 22 years old, he set off on a 5-year voyage around the world on H.M.S *Beagle*. On this voyage, Darwin observed a multitude of living things as well as the remains of ancient organisms. His observations inspired him to solve the riddles of life's diversity.

Where did the different kinds of living things come from? Why have many kinds of living things vanished over time? Why are modern organisms similar to—and different from—ancient organisms? Why are organisms so well designed to do the things they need to do to survive? Many years after his voyage on H.M.S. *Beagle*, Darwin proposed answers to these questions in his book, *On the Origin of Species by Means of Natural Selection*.

In *On the Origin of Species*, Darwin maintained that modern organisms have been produced through evolution from ancient organisms and all species have common ancestors. Darwin also proposed a mechanism to explain exactly how and why evolution occurs.

### Matching Terms: Building Vocabulary Skills

In the space provided, write the term that best fits each definition. Some terms may be used more than once.

_____ **1.** Process that results in fitness

_____ **2.** The combination of physical traits and behaviors that help organisms survive and reproduce

_____ **3.** Process by which modern organisms have descended from ancient organisms

_____ **4.** Person who contributed the most to our understanding of evolution

_____ **5.** Principle that species have shared ancestors

_____ **6.** An inherited characteristic that increases an organism's fitness

_____ **7.** The ship on which Darwin took his famous voyage around the world

_____ **8.** Change over time

_____ **9.** Author of *On the Origin of Species by Means of Natural Selection*

_____ **10.** A characteristic that enhances an organism's chances of passing on its genes

### Observing Adaptations: Using the Main Ideas

Explain how each of the following adaptations improves the organism's fitness.

1. Because of their shape and coloring, sargassum fish resemble certain seaweeds.

_____

_____

_____

_____

_____

_____

_____

2. A flower's fragrance attracts bees.

_____

_____

_____

_____

_____

_____

_____

_____

_____

### Concept Mapping

The construction of and theory behind concept mapping are discussed on pages vii–ix in the front of this Study Guide. Read those pages carefully. Then consider the concepts presented in Section 13–1 and how you would organize them into a concept map. Now look at the concept map for Chapter 13 on page 134. Notice that the concept map has been started for you. Add the key facts and concepts you feel are important for Section 13–1. When you have finished the chapter, you will have a completed concept map.

**The Age of the Earth** *(pages 272–277)*

## SECTION REVIEW

There is an enormous amount of evidence that proves that evolution has occurred. Much of this evidence is found in the rocks of the Earth itself. In this section you read about the geological evidence that shows that the Earth has been around for a very long time and has changed over time.

The ideas of James Hutton and Charles Lyell were particularly important in shaping our view of the Earth. Both men maintained that the Earth is ancient and has been shaped by the same natural geological forces that are still at work. Lyell also argued that scientists must always explain past events in terms of events and processes they can observe themselves.

At the same time that Hutton and Lyell were proposing their theories, other geologists—pro-fessional and amateur—were making some startling discoveries. They found fossils in some of the rocks they were examining and recognized the fossils for what they were: preserved remains of ancient organisms.

The approximate age of fossils can be determined through relative dating. This technique is based on the fact that new layers of rock are deposited on top of older layers of rock. The actual age of fossils can be determined through radioactive dating, which is also known as absolute dating.

Information from relative and radioactive dating has been combined to produce a geologic history of the Earth. Through radioactive dating and examination of the rates of geological processes, scientists have determined that the Earth is about 4½ billion years old.

### Seeing Relationships: Building Vocabulary Skills

In the space provided, describe how the paired terms are related to each other.

**1.** Organism, fossil: _____

_____

_____

**2.** Calendar, geologic time scale: _____

_____

_____

_____

**3.** Layer cake, relative dating: _____

_____

_____

_____

_____

_____

**4.** Clock, radioactive decay: _____

_____

_____

_____

**5.** Era, period: _____

_____

_____

**6.** Absolute dating, half-life: _____

_____

_____

_____

### Recognizing Limitations: Understanding the Main Ideas

Both relative and absolute dating are used in the study of fossils. If you were a scientist, you would have to know which method to use in a given situation. The questions below reflect some of the problems that might be encountered in trying to date fossils.

**1.** When would the layers of rocks not be a good indicator of the age of a fossil?

_____

_____

_____

**2.** Could you date a fossil over one million years old using carbon-14 dating? (*Hint:* Look up

the half-life of carbon-14.) Explain your answer. _____

_____

_____

_____

### Concept Mapping

The construction of and theory behind concept mapping are discussed on pages vii–ix in the front of this Study Guide. Read those pages carefully. Then consider the concepts presented in Section 13–2 and how you would organize them into a concept map. Now look at the concept map for Chapter 13 on page 134. Notice that the concept map has been started for you. Add the key facts and concepts you feel are important for Section 13–2. When you have finished the chapter, you will have a completed concept map.

**The Fossil Record** *(pages 278-281)*

## SECTION REVIEW

In this section you learned about the fossil record. You read about some of the ways in which fossils are formed. You were also introduced to some of the ways in which scientists use fossils to learn more about the history of the plants and animals that live on Earth today and those that have lived here in the past.

Most fossils are found in sedimentary rock. These organisms or parts of organisms are often petrified, or turned to rock. Because conditions have to be just right, not all organisms form fossils. When scientists study the fossil record, they have to accept that there may be many gaps because organisms have vanished from the Earth without a trace. In addition, many fossils are incomplete or poorly preserved and thus are difficult to interpret. However, in spite of the chancy nature of fossil formation and preservation, paleontologists have collected millions of fossils that present a preserved collective history of the Earth's organisms. Perhaps the most important thing that can be learned from the fossil record is that change after change has occurred on the Earth.

### Relating Ideas: Building Vocabulary Skills

In the space provided, write three complete sentences about each of the following terms that will tell something important about that word.

**1.** Paleontologist: _____

_____

_____

_____

_____

**2.** Fossil record: _____

_____

_____

_____

_____

### Concept Mapping

The construction of and theory behind concept mapping are discussed on pages vii–ix in the front of this Study Guide. Read those pages carefully. Then consider the concepts presented in Section 13–3 and how you would organize them into a concept map. Now look at the concept map for Chapter 13 on page 134. Notice that the concept map has been started for you. Add the key facts and concepts you feel are important for Section 13–3. When you have finished the chapter, you will have a completed concept map.

**Section 13–4**    **Evidence from Living Organisms**    *(pages 283–285)*

## SECTION REVIEW

In this section you learned that all living organisms carry within their body evidence that shows the ongoing process of evolution. The history that links living organisms with their ancestors is revealed by similarities in early development, in body structure, and in biochemistry that are shared by related organisms.

The embryos of many different animals are so similar that it is difficult to tell them apart. This indicates that similar genes are at work. The genes that control early development are the shared heritage from a common ancestor.

Homologous structures also provide evidence of a shared evolutionary past. Homologous structures develop from the same basic body parts. However, they appear dissimilar because they are adapted to different functions. A human's arm, a bird's wing, a horse's foreleg, and a whale's flipper are homologous structures. In some cases, a vestigial structure in one animal may be homologous to a functional structure in a related animal.

Similarities in the chemical compounds in organisms are also evidence that the organisms are descended from a common ancestor. The more closely related two types of organisms are, the more similar their important chemical compounds.

### Defining Terms: Building Vocabulary Skills

In your own words, define each of the following terms.

**1.** Homology: _____

_____

_____

_____

**2.** Embryo: _____

_____

_____

**3.** Vestigial organ: _____

_____

_____

_____

**4.** Homologous structures: _____

_____

_____

_____

### Examining Homologies: Using the Main Ideas

The accompanying illustration shows the bones in the forelimbs of some vertebrates. These limbs are drawn to approximately the same scale. Study the illustration, then answer the questions that follow.

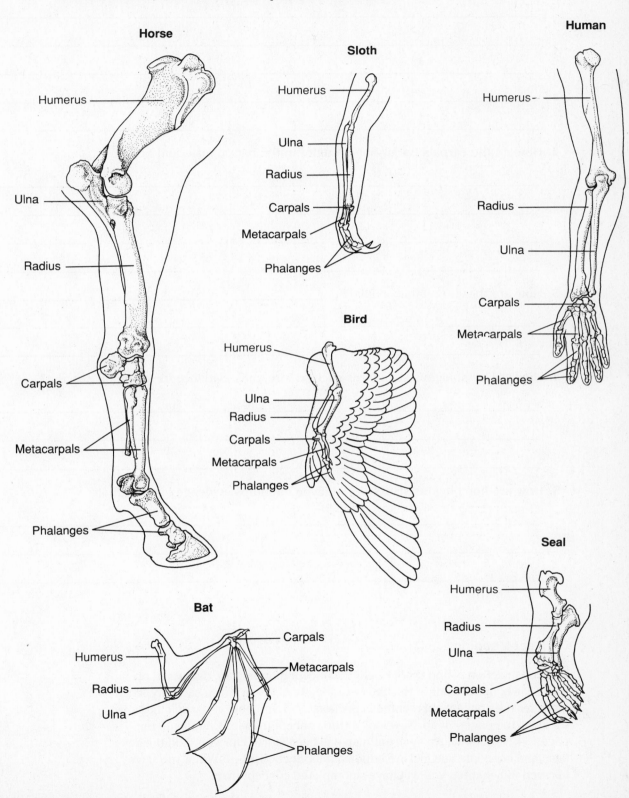

1. How do the phalanges (finger bones) and metacarpals (hand bones) differ in these

   vertebrates? _____

   _____

   _____

   _____

   _____

   _____

2. How do the carpals (wrist bones) differ in the bat, human, and horse?

   _____

   _____

   _____

   _____

3. How are these forelimbs similar? _____

   _____

   _____

4. How are homologous structures such as forelimbs evidence for common descent?

   _____

   _____

   _____

5. How are homologous structures such as forelimbs evidence for evolution?

   _____

   _____

   _____

## Concept Mapping

The construction of and theory behind concept mapping are discussed on
pages vii–ix in the front of this Study Guide. Read those pages carefully. Then
consider the concepts presented in Section 13–4 and how you would organize
them into a concept map. Now look at the concept map for Chapter 13 on
page 134. Notice that the concept map has been started for you. Add the key
facts and concepts you feel are important for Section 13–4. When you have
finished the chapter, you will have a completed concept map.

## Using the Writing Process

Use your writing skills and imagination to respond to the following writing assignment. You will probably need an additional sheet of paper to complete your response.

Imagine that you have been appointed to a committee in charge of selecting science textbooks for your school. Also on the committee are a few people who think that evolution should not be taught because they believe that evolution does not occur. You have been asked by the other committee members to persuade these people to change their opinions.

Prepare a speech in which you present a well-constructed argument in favor of teaching evolution. In your speech, give convincing evidence that evolution has occurred and continues to occur.

_____

_____

_____

_____

_____

_____

_____

_____

_____

_____

_____

_____

_____

_____

_____

_____

_____

_____

_____

## Concept Mapping

The concept map below has been started for you. Add the key facts and concepts for each section of the chapter to this partial concept map. When you are done, you will have a concept map for the entire chapter.

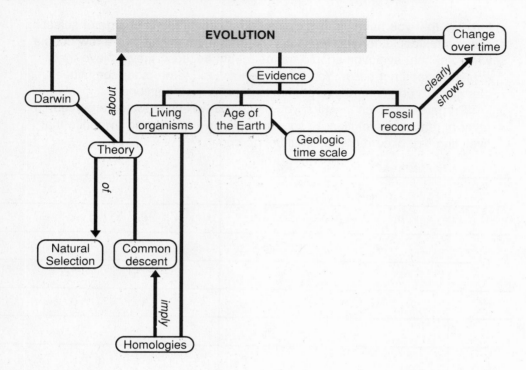

STUDY
GUIDE

| Section 14–1 | Developing a Theory of Evolution | (pages 291–295) |

## SECTION REVIEW

In this section you learned that evolutionary theory is a collection of carefully reasoned hypotheses about how evolutionary change occurs. The fact that plants and animals have changed over time has been obvious to scientists for many years. How and why this change occurs is not so obvious.

Jean Lamarck proposed one of the first explanations for evolutionary change. However, his theory was based on three assumptions that we now know to be incorrect: (1) Organisms have an inborn urge to better themselves, (2) organs can change in size and shape in response to the needs of the organism, and (3) acquired traits can be passed from parents to offspring.

Charles Darwin's explanation for evolutionary change was influenced by the ideas of many other people. After reading Charles Lyell's book, Darwin became convinced that the Earth had existed for a long time—long enough for evolution to occur. The work of plant and animal breeders demonstrated that crops and livestock changed over time as a result of artificial selection. This indicated to Darwin that organisms could—and did—evolve as the result of naturally occurring selective forces. The ideas of Thomas Malthus also had a powerful influence on Darwin. In writing about human population and society, Malthus noted that the animal and plant populations are, like the human population, restricted to a certain size. Darwin recognized that the forces that restrict the growth of populations were also the forces behind evolutionary change.

### Thinking It Through: Building Vocabulary Skills

Recall that the scientific usage of the term *theory* is different from the common usage. In common usage, a theory may be an idea or a hunch that is not backed up by strong evidence. In scientific usage, a theory is a verified, time-tested concept that logically explains past events and makes useful and dependable predictions about the natural world.

Read each of the following statements about evolutionary theory carefully. In the space provided, indicate whether each statement is true or false.

_____ 1. Evolutionary change is undeniable.

_____ 2. Evolutionary theory is based on vague hunches.

_____ 3. Current evolutionary theory is exactly like Darwin's theory.

_____ 4. Much research in many branches of biology is based on evolutionary theory.

_____ 5. Because some of Darwin's original theory was not entirely correct, it is possible that evolution does not occur.

_____ 6. Evolutionary theory is a collection of carefully reasoned hypotheses about how the evolutionary change occurs.

_____ **7.** Evolutionary theory is the foundation on which the rest of biological science is built.

_____ **8.** Darwin was the first person to come up with an explanation for evolution.

### Influences on Darwin: Examining the Main Ideas

1. People who read Darwin's book *On the Origin of Species by Means of Natural Selection* are often surprised that many pages at the beginning of the book discuss the breeding of pigeons. In these pages, Darwin also described the many varieties of domesticated pigeons. Why do you think Darwin chose to spend so much time discussing pigeons?

_____

_____

_____

_____

_____

_____

_____

_____

2. The following passage is from the first chapter of *An Essay on the Principle of Population as It Affects the Future Improvement of Society,* by Thomas Malthus.

> Through the animal and vegetable kingdoms, nature has scattered the seeds of life abroad with the most profuse and liberal hand. She has been comparatively sparing in the room, and the nourishment necessary to rear them. The germs of existence contained in this spot of earth, with ample food, and ample room to expand it, would fill millions of worlds in the course of a few thousand years. Necessity, that imperious, all-pervading law of nature, restrains them within the prescribed bounds. The race of plants, and the race of animals shrink under this great restrictive law. And the race of man cannot, by any efforts of reason, escape from it. Among plants and animals its effects are waste of seed, sickness, and premature death. ...

**a.** What are the main ideas in this passage? _____

_____

_____

_____

_____

_____

_____

**b.** In your own words, define the "great restrictive law" mentioned in the passage.

_____

_____

_____

_____

_____

**c.** According to this passage, how does nature keep the growth of a plant or an animal

population under control?_____

_____

_____

_____

_____

**d.** How might the forces that control the size of populations relate to selection?

_____

_____

_____

_____

_____

### Concept Mapping

The construction of and theory behind concept mapping are discussed on pages vii–ix in the front of this Study Guide. Read those pages carefully. Then consider the concepts presented in Section 14–1 and how you would organize them into a concept map. Now look at the concept map for Chapter 14 on page 146. Notice that the concept map has been started for you. Add the key facts and concepts you feel are important for Section 14–1. When you have finished the chapter, you will have a completed concept map.

**Evolution by Natural Selection** *(pages 296–298)*

## SECTION REVIEW

In this section you were introduced to the concept of natural selection. Darwin proposed that natural selection operates in a fashion similar to the process of artificial selection used by farmers. Darwin observed that plants and animals showed variations, and he realized that many of them were inherited. He also observed that high birthrates and a shortage of life's necessities forced organisms to constantly struggle in order to exist. Darwin concluded that those organisms that possessed characteristics that made them well-suited for their environment survived; those less well-equipped did not survive. Darwin called this principle *survival of the fittest.*

In this section you also read about an interesting example of natural selection in action. Recall that peppered moths come in two forms: a lighter colored form and a darker colored form. The form that blends in with the tree bark in a given area is more common than the noncamouflaged form. An experiment by H.B.D. Kettlewell demonstrated that the camouflaged moth form has a higher survival rate.

### Writing Definitions: Building Vocabulary Skills

In your own words, define each of the following terms.

1. Natural selection: _____

_____

_____

_____

2. Survival of the fittest: _____

_____

_____

_____

3. H.B.D. Kettlewell: _____

_____

_____

_____

### Concept Mapping

The construction of and theory behind concept mapping are discussed on pages vii–ix in the front of this Study Guide. Read those pages carefully. Then consider the concepts presented in Section 14–2 and how you would organize them into a concept map. Now look at the concept map for Chapter 14 on page 146. Notice that the concept map has been started for you. Add the key facts and concepts you feel are important for Section 14–2. When you have finished the chapter, you will have a completed concept map.

## SECTION REVIEW

Darwin worked under a serious handicap when he was developing his theory of evolution: He did not know how inheritable traits are passed from parents to offspring. And as you learned in this section, genetics and evolutionary theory are inseparable.

In this section you discovered how evolutionary theory is defined in genetic terms. You also became familiar with the terms that link genetics and evolution. For example, *genes* are the source of the variation upon which natural selection operates, and the group of genes possessed by a population is known as its *gene pool.* In addition, you learned new definitions for familiar terms. Evolution can be defined as any change in the relative frequency of alleles in the gene pool of a population. Species can be defined as a group of similar-looking organisms that breed with one another and produce fertile offspring in the natural environment.

### Matching Definitions: Building Vocabulary Skills

In the space provided, write the term that best fits each of the following descriptions.

_____   **1.** All the behavioral and physical characteristics produced by the interaction of genotype and environment

_____   **2.** Any change in the relative frequencies of alleles in a gene pool

_____   **3.** The success of an organism in passing on its genes

_____   **4.** A collection of individuals of the same species in a given area whose members can breed with one another

_____   **5.** A graph of this often forms a bell-shaped curve

_____   **6.** The number of times an allele occurs in a gene pool compared with the number of times other alleles for the same gene occur

_____   **7.** Any genetically controlled characteristic of an organism that increases its fitness

_____   **8.** Once defined as a group of organisms that look alike

### Concept Mapping

The construction of and theory behind concept mapping are discussed on pages vii–ix in the front of this Study Guide. Read those pages carefully. Then consider the concepts presented in Section 14–3 and how you would organize them into a concept map. Now look at the concept map for Chapter 14 on page 146. Notice that the concept map has been started for you. Add the key facts and concepts you feel are important for Section 14–3. When you have finished the chapter, you will have a completed concept map.

**The Development of New Species**    *(pages 304–310)*

## SECTION REVIEW

In the first part of this section you were introduced to the concept of the niche. A species's niche is the combination of its role in the environment and the place in which it lives. If two species occupy two different niches, they can coexist. If their niches are the same or extremely similar, they will compete with each other for food and living space. If one species is much better at making a living than the other, it may cause its competitor to become extinct. No two species can occupy the same niche in the same location for a long period of time.

You then learned about speciation, the process in which new species evolve from old species, by reading about Darwin's finches. Recall that the fourteen species of Darwin's finches evolved from a single ancestral species. In the past, populations of the ancestral finches became separated from other populations. The separated populations adapted to local conditions. In time, the gene pools of the finch populations had changed so much that the finch populations could not interbreed.

Finally, you learned about some basic patterns of evolution. Adaptive radiation, or divergent evolution, is the process in which an ancestral species gives rise to a diversity of new species. Convergent evolution is the process in which species from different evolutionary lines grow to resemble one another because they adapt to similar conditions.

### Defining Terms: Building Vocabulary Skills

In your own words, define each of the following terms.

**1.** Divergent evolution: _____

_____

**2.** Analogous structures: _____

_____

**3.** Niche: _____

_____

**4.** Reproductive isolation: _____

_____

**5.** Adaptive radiation: _____

_____

**6.** Speciation: _____

_____

**7.** Convergent evolution: _____

_____

### Analyzing Diagrams: Using the Main Ideas

Some of the animals shown in the accompanying diagrams lived long ago. Others are still alive today. Examine diagrams A and B. Then answer the following questions.

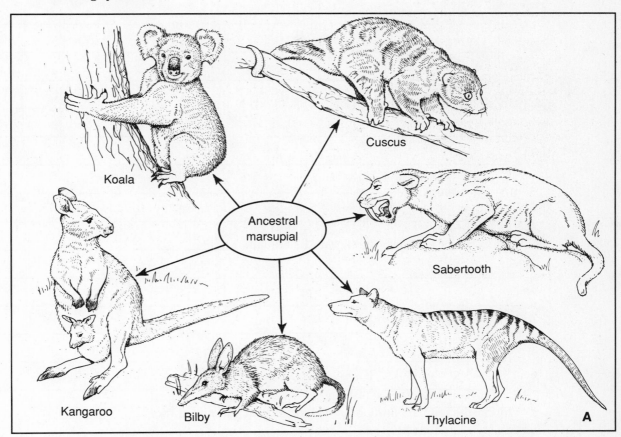

Koala

Cuscus

Sabertooth

Ancestral marsupial

Kangaroo

Bilby

Thylacine

A

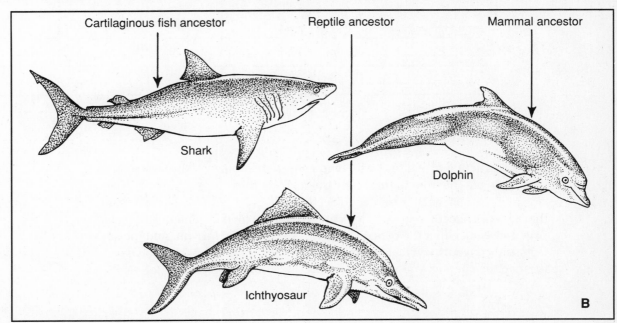

Cartilaginous fish ancestor

Reptile ancestor

Mammal ancestor

Shark

Dolphin

Ichthyosaur

B

**1.** What process is shown in diagram A? Explain. _____

_____

_____

_____

_____

_____

**2.** Why do the animals in diagram A look different from one another? _____

_____

_____

_____

_____

_____

**3.** What process is shown in diagram B? Explain. _____

_____

_____

_____

_____

_____

**4.** Why do the animals in diagram B look similar to one another? _____

_____

_____

_____

_____

_____

### Concept Mapping

The construction of and theory behind concept mapping are discussed on
pages vii–ix in the front of this Study Guide. Read those pages carefully. Then
consider the concepts presented in Section 14–4 and how you would organize
them into a concept map. Now look at the concept map for Chapter 14 on
page 146. Notice that the concept map has been started for you. Add the key
facts and concepts you feel are important for Section 14–4. When you have
finished the chapter, you will have a completed concept map.

## SECTION REVIEW

Evolutionary theory has been modified over the years as scientists formulate theories about the details of evolutionary change. In this section you learned about some of the major modifications to Darwin's original theory of evolution by natural selection.

Sometimes the relative frequency of an allele in a gene pool changes merely by chance. In other words, evolution may occur in the absence of natural selection. This is known as genetic drift.

Darwin envisioned evolution as being a slow, gradual, and continuous process. However, this is not always the case. In a few cases, evolution may occur so slowly that gene pools remain the same for long periods of time. Alternatively, evolution may occur relatively rapidly.

### Defining Terms: Building Vocabulary Skills

In your own words, define each of the following terms.

**1.** Punctuated equilibria:_____

_____

_____

_____

**2.** Genetic drift: _____

_____

_____

_____

**3.** Gradualism: _____

_____

_____

_____

**4.** Mass extinction: _____

_____

_____

_____

**5.** Equilibrium:_____

_____

_____

### Understanding Concepts: Using the Main Ideas

1. Under what circumstances might genetic drift occur? _____

   _____

   _____

   _____

   _____

   _____

2. What does genetic drift imply about an organism's characteristics and its fitness?

   _____

   _____

   _____

   _____

3. How might the theory of punctuated equilibria explain "missing links" in the fossil

   record? _____

   _____

   _____

   _____

   _____

4. Under what circumstances might rapid evolution occur? _____

   _____

   _____

   _____

   _____

   _____

### Concept Mapping

The construction of and theory behind concept mapping are discussed on
pages vii–ix in the front of this Study Guide. Read those pages carefully. Then
consider the concepts presented in Section 14–5 and how you would organize
them into a concept map. Now look at the concept map for Chapter 14 on
page 146. Notice that the concept map has been started for you. Add the key
facts and concepts you feel are important for Section 14–5. When you have
finished the chapter, you will have a completed concept map.

## Using the Writing Process

Use your writing skills and imagination to respond to the following writing assignment. You will probably need an additional sheet of paper to complete your response.

The evolutionary success of a group of organisms is generally judged by factors such as the number of species it contains, the number of niches it occupies, and the size of its geographic range. In the past, many groups of organisms were considered to be less fit or somehow inferior because they were not successful. However, some experts now think that a group's success—or failure—may be credited as much to luck as to fitness.

Following a mass extinction about 65 million years ago, mammals took over many of the niches formerly inhabited by dinosaurs and became the dominant group of animals. But it is possible that a different group of animals might have become dominant. Write a short story or essay that describes what life on Earth might be like today if birds had become the dominant vertebrates.

_____

_____

_____

_____

_____

_____

_____

_____

_____

_____

_____

_____

_____

_____

_____

_____

_____

_____

## Concept Mapping

The concept map below has been started for you. Add the key facts and concepts for each section of the chapter to this partial concept map. When you are done, you will have a concept map for the entire chapter.

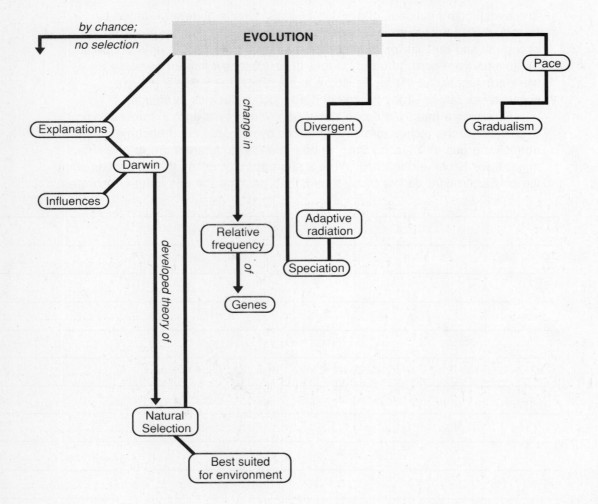

S T U D Y
G U I D E

| Section 15–1 | **Why Classify?** | *(pages 319–320)* |
| --- | --- | --- |

**SECTION REVIEW**

In this section you discovered that there are more than 2 ¹/₂ million species of organisms on Earth. In order to study and understand this great diversity of organisms, scientists must name them and divide them into categories in a logical manner. In other words, they must classify them. Good classification systems have two important characteristics: (1) They are universally accepted, and (2) they place organisms into groups that have real biological meaning.

**Making the Rules: Understanding the Main Ideas**

Explain why each of the following characteristics of a classification system is important.

**1.** It assigns a single universally accepted name to each organism. _____

_____

_____

_____

**2.** It places organisms into groups that have real biological meaning. _____

_____

_____

_____

**3.** It divides organisms into small groups. _____

_____

_____

_____

**Concept Mapping**

The construction of and theory behind concept mapping are discussed on pages vii–ix in the front of this Study Guide. Read those pages carefully. Then consider the concepts presented in Section 15–1 and how you would organize them into a concept map. Now look at the concept map for Chapter 15 on page 154. Notice that the concept map has been started for you. Add the key facts and concepts you feel are important for Section 15–1. When you have finished the chapter, you will have a completed concept map.

| Section 15–2 | Biological Classification | *(pages 320 – 323)* |

## SECTION REVIEW

About 200 years ago, European scientists stopped using common names in local languages to describe organisms. Instead, they started a practice that is continued today: They used names based on Latin or ancient Greek words because those languages were understood by educated people everywhere. Unfortunately, many of the early naming systems were quite cumbersome and difficult to standardize.

As you learned in this section, the Swedish botanist Carolus Linnaeus devised a system for naming organisms. This system, known as binomial nomenclature, soon gained wide acceptance and is still in use today.

In addition to giving each organism a two-part scientific name, Linnaeus placed organisms into groups based on shared body structures. These groups are known as taxa. The taxa are, in order from largest to smallest, kingdom, phylum, class, order, family, genus, species.

### Completing Sentences: Building Vocabulary Skills

In the space provided, write the term that best fits each of the following sentences.

1. In _____, an organism is given a two-part scientific name that gives the organism's genus and species.

2. The science of naming organisms and putting them into classification groups is known as

   _____.

3. _____ devised a system of naming organisms that is still in use today.

4. Organisms are placed in _____, or classification groups.

5. The taxon that is larger than a genus and smaller than an order is a (an)

   _____.

6. The smallest taxon is the _____, which is made up of organisms that share similar characteristics and can breed with one another.

### Devising Mnemonics: Remembering the Main Ideas

A mnemonic (nee-MAHN-ihk) device is a rhyme or phrase that helps you remember an important date, fact, or list of items. One that you may be familiar with is, "In 1492, Columbus sailed the ocean blue." A mnemonic device for taxonomy used by generations of biology students is, "**K**ings **p**lay **c**hess **o**n **f**ine **g**reen **s**and." The words in this nonsensical sentence begin with the same letters of the taxa listed from largest to smallest.

In the space provided, invent two mnemonics of your own for the seven taxa.

1. _____

   _____

2. _____

_____

### Putting It Together: Using the Main Ideas

Using the information in this section and elsewhere in the textbook, complete the accompanying table. (*Hint:* Use the index in the back of the textbook to help you find information.)

| Classification of the Common House Cat | |
| --- | --- |
| **Taxon** | **Taxon Characteristics** |
| Kingdom: | |
| Phylum: | |
| Class: | |
| Order: | |
| Family: | |
| Genus: | |
| Species: | |

### Concept Mapping

The construction of and theory behind concept mapping are discussed on pages vii–ix in the front of this Study Guide. Read those pages carefully. Then consider the concepts presented in Section 15–2 and how you would organize them into a concept map. Now look at the concept map for Chapter 15 on page 154. Notice that the concept map has been started for you. Add the key facts and concepts you feel are important for Section 15–2. When you have finished the chapter, you will have a completed concept map.

| Section 15–3 | **Taxonomy Today** | *(pages 323–325)* |

## SECTION REVIEW

In this section you discovered that taxonomy is not as simple as it seems. Although the taxa above the level of species are useful in organizing information about living things, they do not have a clear biological identity. Thus it is difficult to draw the lines between groups. Taxonomists now try to group organisms in ways that reflect their evolutionary relationships. This means that they must examine similarities among organisms more closely and distinguish between analogous and homologous structures. Unlike the taxonomists of Linnaeus's time, modern taxonomists study more than the visible characteristics of organisms. They also examine the chemical compounds in living things. Similarities and differences in these compounds provide clues about evolutionary relationships.

### Explaining Concepts: Understanding the Main Ideas

1. Explain why taxonomists may not agree on the classification of an organism.

   _____

   _____

   _____

   _____

   _____

2. List three sources of information that taxonomists use in determining evolutionary

   relationships between groups of organisms. _____

   _____

   _____

   _____

   _____

   _____

### Concept Mapping

The construction of and theory behind concept mapping are discussed on pages vii–ix in the front of this Study Guide. Read those pages carefully. Then consider the concepts presented in Section 15–3 and how you would organize them into a concept map. Now look at the concept map for Chapter 15 on page 154. Notice that the concept map has been started for you. Add the key facts and concepts you feel are important for Section 15–3. When you have finished the chapter, you will have a completed concept map.

**Section 15–4**  **The Five-Kingdom System**  *(pages 325–329)*

## SECTION REVIEW

As biologists accumulated more information about living things in the centuries after Linnaeus devised his classification system, it became clear that more than two kingdoms were needed to categorize organisms logically. Today, the most generally accepted classification system contains five kingdoms: Monera, Protista, Fungi, Plantae, and Animalia. Prokaryotes are placed in the kingdom Monera. Single-celled eukaryotes belong to the kingdom Protista. (However, a few single-celled eukaryotes are often grouped with fungi or plants.) The kingdom Fungi contains heterotrophic eukaryotes whose cell walls do not contain cellulose and may not completely separate the individual cells. The kingdom Plantae contains multicellular photosynthetic autotrophs whose cell walls contain cellulose. Multicellular heterotrophs whose cells lack cell walls are placed in the kingdom Animalia.

### Classifying Organisms: Building Vocabulary Skills

Examine the accompanying illustrations and descriptions of organisms. Underline the information in the description that allows you to identify the kingdom to which each organism belongs. In the space provided, write the name of the kingdom.

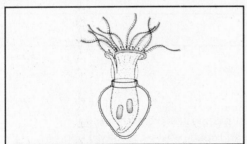

1. Musk deer: Endangered; lives in Asia; lacks cell walls; eats plants; produces musk, which is used in perfumes; males have long canine teeth

   Kingdom: _____

2. *Tintinnopsis:* Microscopic; feeds with brushes made up of fused cilia; unicellular; surrounded by a jellylike coat; possesses two large nuclei

   Kingdom: _____

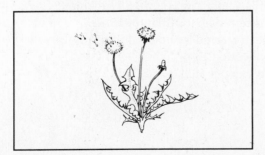

3. Dandelion: Cell walls contain cellulose; seeds dispersed by the wind; photosynthetic; yellow flowers; made up of many cells; eukaryote

   Kingdom: _____

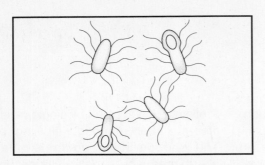

**4.** *Clostridium tetani:* Rod-shaped; internal spores make some cells drumstick-shaped; lacks a nucleus; produces a toxin that causes tetanus (lockjaw)

Kingdom: _____

**5.** Morel: Up to 10–15 centimeters high; blackish gray in color; spongelike in appearance; good to eat; cell walls contain chitin rather than cellulose

Kingdom: _____

### Comparing Kingdoms: Using the Main Ideas

Use the information in Section 15–4 to complete the following table.

| Kingdom | Characteristics | Example |
|---------|-----------------|---------|
| Animalia | | |
| Fungi | | |
| Monera | | |
| Plantae | | |
| Protista | | |

### Concept Mapping

The construction of and theory behind concept mapping are discussed on pages vii–ix in the front of this Study Guide. Read those pages carefully. Then consider the concepts presented in Section 15–4 and how you would organize them into a concept map. Now look at the concept map for Chapter 15 on page 154. Notice that the concept map has been started for you. Add the key facts and concepts you feel are important for Section 15–4. When you have finished the chapter, you will have a completed concept map.

## Using the Writing Process

Use your writing skills and imagination to respond to the following writing assignment. You will probably need an additional sheet of paper to complete your response.

Imagine that you are one of the first biologists to study the living things on a distant planet. Your job is to develop a preliminary classification scheme so that these organisms can be studied in a systematic way.

Prepare a notebook of your first week or two on the planet. Describe and sketch at least ten of the organisms you observe. Note how these organisms compare with those on Earth. Discuss how these organisms might be classified. Describe any problems or concerns you have about your assignment.

_____

_____

_____

_____

_____

_____

_____

_____

_____

_____

_____

_____

_____

_____

_____

_____

_____

_____

## Concept Mapping

The concept map below has been started for you. Add the key facts and concepts for each section of the chapter to this partial concept map. When you are done, you will have a concept map for the entire chapter.

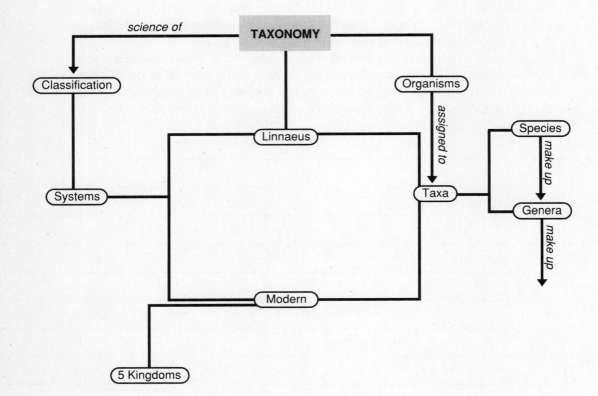

S T U D Y
G U I D E

**Section 16–1** **Spontaneous Generation** *(pages 339–342)*

### SECTION REVIEW

In this section you examined the hypothesis of spontaneous generation. Spontaneous generation refers to the idea that life regularly arises from nonlife, or nonliving things. You looked into the experiments of Francesco Redi and Lazzaro Spallanzani, which attempted to disprove the spontaneous generation hypothesis. You also examined the experiments of John Needham, which tried to prove that the spontaneous generation hypothesis was correct. Finally, you studied the experiments of Louis Pasteur, which showed conclusively that the spontaneous generation hypothesis was incorrect.

### Spontaneous Generation: Relating Facts

Describe in you own words the experiments of each of the following scientists and whether the experiment proved or disproved the spontaneous generation hypothesis.

Francesco Redi: _____

_____

_____

_____

Lazzaro Spallanzani: _____

_____

_____

_____

John Needham: _____

_____

_____

_____

Louis Pasteur: _____

_____

_____

_____

### Scrambled Experiments: Finding the Main Ideas

The illustrations below show the steps in the experiments of Redi, Spallanzani, and Pasteur. Unfortunately, they are scrambled and out of order. Unscramble the steps. Then place the correct order of steps beside the names of each of the scientists listed below. Use the letter key beside each step to mark the correct order.

Redi: _____

Spallanzani: _____

Pasteur: _____

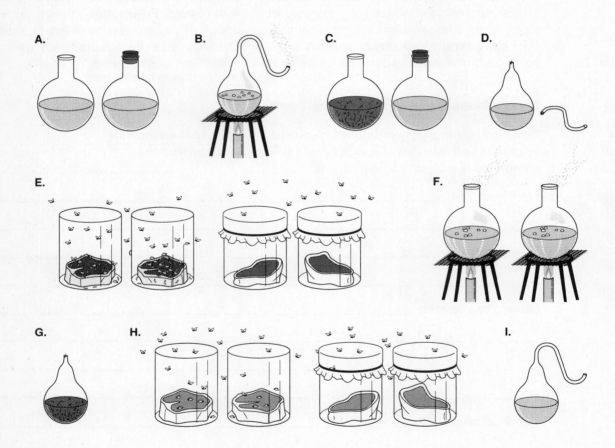

A.    B.    C.    D.

E.    F.

G.    H.    I.

### Concept Mapping

The construction of and theory behind concept mapping are discussed on pages vii–ix in the front of this Study Guide. Read those pages carefully. Then consider the concepts presented in Section 16–1 and how you would organize them into a concept map. Now look at the concept map for Chapter 16 on page 160. Notice that the concept map has been started for you. Add the key facts and concepts you feel are important for Section 16–1. When you have finished the chapter, you will have a completed concept map.

**Section 16–2** | **The First Signs of Life** | *(pages 342–346)*

## SECTION REVIEW

In this section you looked into some of the theories as to how life arose on Earth. You discovered that organic molecules such as amino acids, lipids, and carbohydrates will form spontaneously during laboratory experiments in which conditions on ancient Earth are simulated.

Scientists have found microfossils that have been dated back to 3.5 billion years ago. The microfossils provide the outlines of ancient cells and show that these first cells were heterotrophic prokaryotes similar to bacteria alive today.

Over time, cells that could perform an ancient form of photosynthesis evolved. These cells were autotrophic because they could produce their own food. The first autotrophs were extremely successful and were commonplace some 3.4 billion years ago.

### Early Cells: Relating the Main Ideas

Place a Y next to each of the following characteristics that relate to the first ancient cells. Place an N next to each characteristic that does not relate to the first cells. In each case, explain your response.

**1.** Contain a nucleus: _____

_____

_____

**2.** Aerobic: _____

_____

_____

**3.** Multicellular: _____

_____

**4.** Photosynthetic: _____

_____

**5.** Prokaryotic: _____

_____

### Concept Mapping

The construction of and theory behind concept mapping are discussed on pages vii–ix in the front of this Study Guide. Read those pages carefully. Then consider the concepts presented in Section 16–2 and how you would organize them into a concept map. Now look at the concept map for Chapter 16 on page 160. Notice that the concept map has been started for you. Add the key facts and concepts you feel are important for Section 16–2. When you have finished the chapter, you will have a completed concept map.

## The Road to Modern Organisms
*(pages 347–349)*

### SECTION REVIEW

In this section you looked into the evolution of life after photosynthetic organisms appeared. You learned that the oxygen released during photosynthesis dramatically changed the composition of the atmosphere and led to our modern atmosphere, which is about one-fifth oxygen. The addition of oxygen to the atmosphere had several important effects. First, it banished anaerobic organisms to parts of the Earth where oxygen can no longer reach. Second, it formed an ozone layer in the upper atmosphere that acts as a shield from the damaging ultraviolet rays of the sun. Third, it allowed the evolution of metabolic pathways that utilize oxygen (aerobic metabolism).

You also discovered that eukaryotic cells evolved around 1.4 billion years ago. Eukaryotic cells contain a nucleus and membrane-bound organelles (mitochondria and chloroplasts, for example). Finally you looked into the evolution of sexual reproduction—which allows for much more genetic variety than asexual reproduction—and the eventual evolution of multicellular life.

### Compare and Contrast: Finding the Main Ideas

For each of the following pairs of characteristics, underline the characteristic that evolved first. Then explain the reason for your selection.

**1.** Autotroph, heterotroph: _____

_____

**2.** Prokaryotic, eukaryotic: _____

_____

**3.** Aerobic metabolism, anaerobic metabolism: _____

_____

**4.** Sexual reproduction, asexual reproduction: _____

_____

**5.** Unicellular, multicellular: _____

_____

### Concept Mapping

The construction of and theory behind concept mapping are discussed on pages vii–ix in the front of this Study Guide. Read those pages carefully. Then consider the concepts presented in Section 16–3 and how you would organize them into a concept map. Now look at the concept map for Chapter 16 on page 160. Notice that the concept map has been started for you. Add the key facts and concepts you feel are important for Section 16–3. When you have finished the chapter, you will have a completed concept map.

## Using the Writing Process

Use your writing skills and imagination to respond to the following writing assignment. You will probably need an additional sheet of paper to complete your response.

Pretend that you are a reporter during the time of Louis Pasteur. Dr. Pasteur has just completed his famous experiment disproving spontaneous generation. However, you are still skeptical. (In your day almost everyone still believes that spontaneous generation is an observable fact.) Write the interview as it might have taken place, providing both questions and Pasteur's responses. Keep in mind the fact that you are limited to the information that was available to people of that time period.

_____

_____

_____

_____

_____

_____

_____

_____

_____

_____

_____

_____

_____

_____

_____

_____

_____

_____

## Concept Mapping

The concept map below has been started for you. Add the key facts and concepts for each section of the chapter to this partial concept map. When you are done, you will have a concept map for the entire chapter.

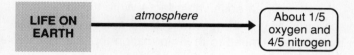

**LIFE ON EARTH** —*atmosphere*→ About 1/5 oxygen and 4/5 nitrogen

**ANCIENT EARTH** —*atmosphere*→ Water vapor, carbon monoxide, carbon dioxide, nitrogen, hydrogen sulfide, hydrogen cyanide

S T U D Y
G U I D E

| Section 17–1 | **Viruses** | *(pages 355– 360)* |

### SECTION REVIEW

In this section you were introduced to the term *virus,* which is derived from the Latin word that means poison. You discovered that the structure of a typical virus is made up of a protein coat called the capsid and a core that contains nucleic acids (DNA or RNA). The core of a virus may contain DNA or RNA but never both. Bacteriophages, or viruses that infect bacteria, are made up of a head region containing the capsid and core, as well as a tail. The tail fibers are used by the virus to attach itself to a bacterium.

In order to reproduce, a virus must invade, or infect, a living cell in a host organism. Typically, a virus becomes attached to a host cell. The nucleic acids (DNA or RNA) within the viral core are injected into the host cell. What happens next depends on the type of virus and its life cycle. In a lytic infection, the nucleic acid from the virus takes over the host cell and uses the materials of the host cell to make thousands of copies of its own protein coat and nucleic acid. Soon the host cell is filled with viruses and the host cell bursts, or lyses, releasing the viruses to infect other cells in the host.

In a lysogenic infection, the virus does not reproduce immediately after infecting the host cell. Instead, the nucleic acid of the virus is inserted into the DNA of the host cell. The viral DNA may stay within the host DNA for quite some time. However, eventually it may become active, remove itself from the host DNA, and begin to direct the production of new viruses. As with a lytic infection, the viruses from a lysogenic infection will burst out of the host cell and are then freed to infect other cells.

### Applying Definitions: Building Vocabulary Skills

Examine the accompanying illustration of a typical bacteriophage (virus that infects bacteria). In the lines provided, identify each part of the virus. Then, in your own words, write a definition for that part.

a. _____

_____

b. _____

_____

c. _____

_____

d. _____

_____

e. _____

_____

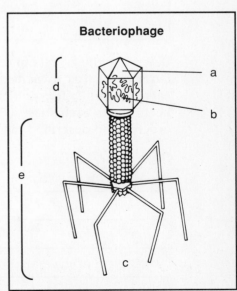

**Bacteriophage**

**Sequencing Events: Finding the Main Ideas**

The stages in the life cycle of a lytic virus are scrambled in the accompanying illustration. Examine the illustration carefully. Complete the key by filling in the name of the structure on the line next to each illustration in the key.

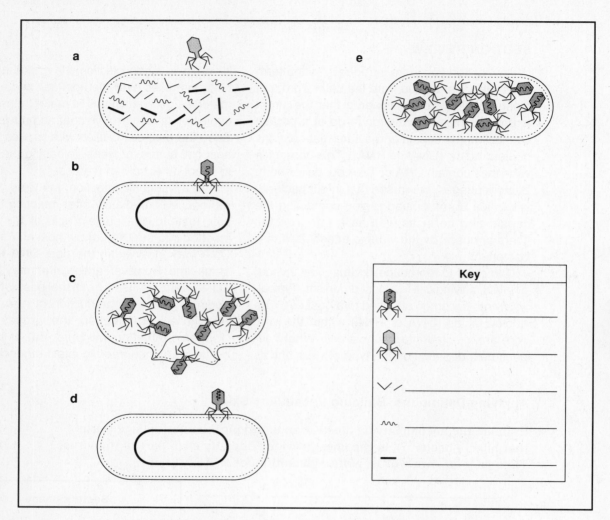

Determine the correct order of the stages in the illustration. Put the letter of the first stage next to the line labeled Stage One. Follow the same procedure for each of the other four stages. Then describe what is occurring in each stage. When you are done, you will have correctly ordered and described the stages of a lytic infection.

**Stage One** ( _____ ) _____

_____

_____

_____

_____

**Stage Two** ( _____ ) _____

_____

_____

_____

_____

**Stage Three** ( _____ ) _____

_____

_____

_____

_____

**Stage Four** ( _____ ) _____

_____

_____

_____

_____

**Stage Five** ( _____ ) _____

_____

_____

_____

_____

### Concept Mapping

The construction of and theory behind concept mapping are discussed on pages vii–ix in the front of this Study Guide. Read those pages carefully. Then consider the concepts presented in Section 17–1 and how you would organize them into a concept map. Now look at the concept map for Chapter 17 on page 168. Notice that the concept map has been started for you. Add the key facts and concepts you feel are important for Section 17–1. When you have finished the chapter, you will have a completed concept map.

## SECTION REVIEW

In this section you were introduced to the bacteria, the single-celled prokaryotic organisms that make up the kingdom Monera. Bacteria are divided into four phyla: Eubacteria, Cyanobacteria, Archaebacteria, and Prochlorobacteria. Bacteria are often identified by their shape. Bacteria can also be identified by the way in which their cell wall is colored by Gram staining.

Bacteria obtain energy in a variety of ways. Some bacteria are autotrophs, which use a source of energy such as sunlight or chemicals to produce food (organic molecules) from simple inorganic molecules. Other bacteria are heterotrophs, which cannot make their food and instead obtain energy from the organic molecules that they eat.

The energy stored in food molecules is made available for use when the food molecules are broken down. Bacteria use the processes of respiration and fermentation to release the energy in food. Because bacteria release energy in various ways, their need for and tolerance of oxygen also varies.

Bacteria reproduce by binary fission. Some bacteria also undergo other processes, such as conjugation and spore formation. During conjugation, genetic material is transferred from one bacterium to another, thus creating new combinations of genes. During one type of spore formation, a bacterium encloses its DNA with a thick internal wall to form a spore. The spore can survive harsh conditions that would kill the bacterium in its active form.

Bacteria fit into the world in many ways. For example, some bacteria help animals digest their food. Some are used in food production and industry. Others convert nitrogen gas into a form that can be used by plants. Still others break down dead material and thus help recycle nutrients in the environment.

### Finding the Oddball: Building Vocabulary Skills

For each of the following sets of terms, determine the characteristic common to three of the terms. Then identify the term that does not belong.

**1.** spirillum, coccus, methanogen, bacillus: _____

_____

_____

**2.** prokaryote, nitrogen fixation, saprophyte, symbiosis: _____

_____

_____

**3.** Eubacteria, Cyanobacteria, Archaebacteria, Azotobacteria: _____

_____

_____

**4.** cell wall, nucleus, flagellum, cytoplasm: _____

_____

_____

### Labeling Diagrams: Using the Main Ideas

Label the diagram of a typical bacterium using the following terms: cell membrane, cell wall, cytoplasm, flagellum, genetic material, ribosome. Then answer the questions that follow the diagram.

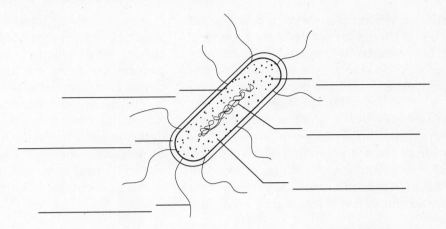

1. Is the bacterium in the diagram a bacillus, coccus, or spirillum? Explain.

_____

2. How would you expect this bacterium to move? _____

3. Suppose that this bacterium was a streptobacillus. What kind of colonies would you

   expect it to form? _____

4. This bacterium is Gram-negative. What happens when it is subjected to Gram staining?

   _____

   _____

5. This bacterium is a facultative aerobe. What process or processes would you expect it to

   use to break down food? Explain. _____

   _____

   _____

   _____

   _____

### Concept Mapping

The construction of and theory behind concept mapping are discussed on pages vii–ix in the front of this Study Guide. Read those pages carefully. Then consider the concepts presented in Section 17–2 and how you would organize them into a concept map. Now look at the concept map for Chapter 17 on page 168. Notice that the concept map has been started for you. Add the key facts and concepts you feel are important for Section 17–2. When you have finished the chapter, you will have a completed concept map.

**Diseases Caused by Viruses and Monerans** *(pages 372–375)*

## SECTION REVIEW

In this section you learned that only a small number of viruses and bacteria are capable of producing disease in humans. You also read about ways in which viral and bacterial diseases can be treated and prevented.

Viruses cause human diseases such as polio, measles, AIDS, and the common cold. Some viruses cause cancers in animals. Many viral diseases are prevented with vaccines. Most viral infections cannot be cured with medicines, although medicines may help treat the symptoms of a viral disease. Scientists are currently studying interferons, proteins produced by virus-infected cells that make it difficult for the viruses to infect other cells. It is possible that many viral diseases may someday be cured using interferons.

Bacteria cause human diseases such as tuberculosis, syphilis, and tetanus. Bacterial diseases can also be treated with drugs such as antibiotics. Bacterial infections can be prevented by controlling bacterial growth through sterilization and proper food processing.

### Defining Terms: Building Vocabulary Skills

In your own words, define each of the following terms.

**1.** Antibiotic: _____

_____

**2.** Disinfectant: _____

_____

**3.** Interferon: _____

_____

**4.** Pathogen: _____

_____

**5.** Sterilization: _____

_____

**6.** Vaccine: _____

_____

### Concept Mapping

The construction of and theory behind concept mapping are discussed on pages vii–ix in the front of this Study Guide. Read those pages carefully. Then consider the concepts presented in Section 17–3 and how you would organize them into a concept map. Now look at the concept map for Chapter 17 on page 168. Notice that the concept map has been started for you. Add the key facts and concepts you feel are important for Section 17–3. When you have finished the chapter, you will have a completed concept map.

## Using the Writing Process

Use your writing skills and imagination to respond to the following writing assignment. You will probably need an additional sheet of paper to complete your response.

Imagine that your best friend desperately needs to pass a biology test on the two types of viral infections. Unfortunately, your friend is completely unable to understand your teacher's explanations or the discussion in the textbook. Knowing that your friend loves to read adventure stories, you have decided to try to present the information about viruses in a new way. Your assignment is to write an adventure story that describes either a lytic or a lysogenic infection.

*Hint:* A lytic infection could be a pirate attacking a merchant ship, a renegade robot taking over a factory, or the invasion of a castle or town. A lysogenic infection could be a "mole"—a spy that infiltrates an enemy's intelligence agency or government. Be creative and have fun!

_____

_____

_____

_____

_____

_____

_____

_____

_____

_____

_____

_____

_____

_____

_____

_____

_____

_____

## Concept Mapping

## Chapter 17

The concept map below has been started for you. Add the key facts and concepts for each section of the chapter to this partial concept map. When you are done, you will have a concept map for the entire chapter.

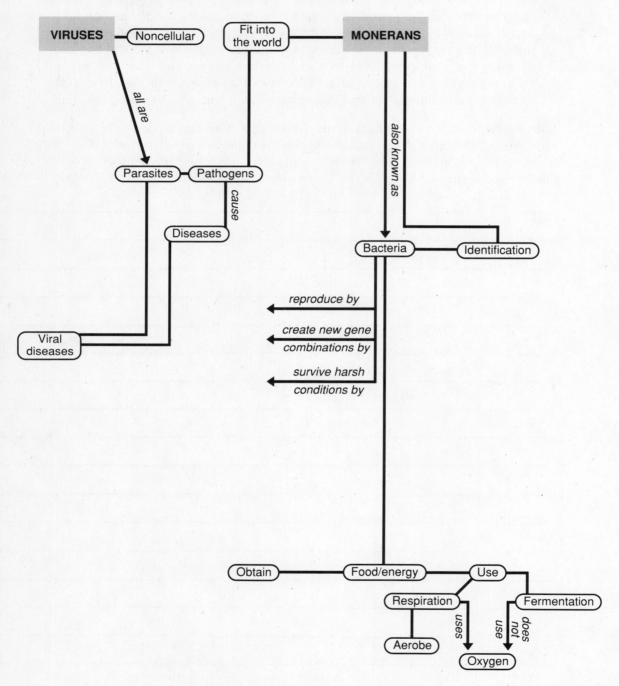

© Prentice-Hall, Inc.

STUDY
GUIDE

| Section 18–1 | **The Kingdom Protista** | *(pages 381–383)* |

### SECTION REVIEW

In this section you were introduced to protists, or members of the kingdom Protista. A protist can be defined as a single-celled eukaryotic organism that is not an animal, plant, or fungus.

Protists, which first appeared about 1.5 billion years ago, were the first group of eukaryotes to evolve. One explanation for the way the first eukaryotes developed from prokaryotes is Lynn Margulis's Endosymbiont Hypothesis. According to this hypothesis, the first eukaryote—and the first protist—was formed by a symbiosis among several prokaryotes. Evidence to support this hypothesis includes structural similarities between certain eukaryotic organelles and bacteria and the bacterial endosymbionts that are found in certain protists alive today.

### Applying Definitions: Building Vocabulary Skills

1. Using your own words, define the word *protist*. _____

_____

_____

2. In the past, why were protists often difficult to classify? _____

_____

_____

3. Explain the Endosymbiont Hypothesis in your own words. _____

_____

_____

4. Describe the evidence that supports the Endosymbiont Hypothesis.

_____

_____

### Concept Mapping

The construction of and theory behind concept mapping are discussed on pages vii–ix in the front of this Study Guide. Read those pages carefully. Then consider the concepts presented in Section 18–1 and how you would organize them into a concept map. Now look at the concept map for Chapter 18 on page 178. Notice that the concept map has been started for you. Add the key facts and concepts you feel are important for Section 18–1. When you have finished the chapter, you will have a completed concept map.

| Section 18–2 | **Animallike Protists** | *(pages 384–394)* |

## SECTION REVIEW

The four phyla of animallike protists were described and discussed in this section. Recall that members of the phyla Ciliophora, Zoomastigina, Sporozoa, and Sarcodina are known as animallike protists.

The members of the phylum Ciliophora are known as ciliates because they have short hairlike structures called cilia. Cilia are used in movement and in feeding. Most ciliates are free-living. Some ciliates are solitary; others are colonial. Ciliates typically have two types of nuclei: a macronucleus and a micronucleus. The macronucleus is the control center of the cell. The micronucleus is involved in conjugation. *Paramecium* is an example of a typical ciliate.

The members of the phylum Zoomastigina move through the water by means of long, whiplike projections called flagella. Some zoomastiginans are free-living. Others, such as trypanosomes, are parasites. Still others, such as *Trichonympha,* are endosymbionts.

The members of phylum Sporozoa are parasitic and nonmotile and reproduce by means of spores. The sporozoan *Plasmodium* causes the human disease malaria. Like many other sporozoans, *Plasmodium* has a life cycle that involves more than one host.

The members of phylum Sarcodina, or sarcodines, possess pseudopods that are used in feeding. Some sarcodines, such as amebas, also use pseudopods in locomotion. There are four groups of sarcodines: amebas, heliozoans, radiolarians, and foraminifers. Some sarcodines produce shells of silica or calcium carbonate. Although most sarcodines are free-living, a few, such as *Entamoeba,* are parasites.

### Identifying Structures: Building Vocabulary Skills

Examine the accompanying diagram of a paramecium. In the spaces provided, identify each part of the paramecium and write a definition for each part.

a. _____

_____

_____

b. _____

_____

_____

c. _____

_____

_____

d. _____

_____

_____

© Prentice-Hall, Inc.

e. _____

_____

_____

f. _____

_____

_____

g. _____

_____

_____

h. _____

_____

_____

i. _____

_____

_____

j. _____

_____

_____

### Discussing Concepts: Finding the Main Ideas

**1.** In your own words, describe the events that occur during conjugation in paramecia.

_____

_____

_____

_____

_____

_____

_____

**2.** Why is conjugation significant? _____

_____

_____

**3.** What are two ways in which animallike protists harm other organisms?

_____

_____

**4.** What are two ways in which animallike protists benefit other organisms?

_____

_____

### Malaria: Interpreting Diagrams

Examine the diagram below of the life cycle of *Plasmodium,* the protist that causes malaria in humans. In the spaces provided on the next page, describe what happens in each of the numbered stages. Two of the stages have been described for you.

### Life Cycle of *Plasmodium*

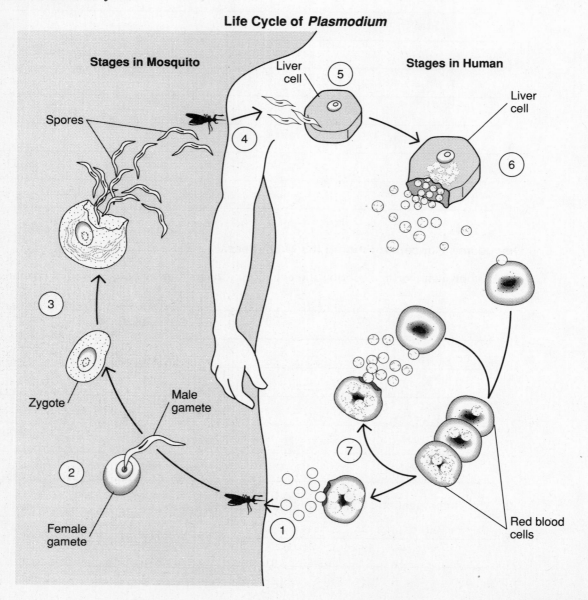

© Prentice-Hall, Inc.

**1.** _____
_____
_____
_____

**2.** *Plasmodium* cells develop into gametes in the mosquito's digestive tract. Fertilization occurs, forming a zygote.

**3.** The zygote attaches to the wall of the mosquito's digestive tract and develops into thousands of spores. Spores migrate to the mosquito's salivary glands.

**4.** _____
_____
_____
_____

**5.** _____
_____
_____
_____

**6.** _____
_____
_____
_____
_____

**7.** _____
_____
_____
_____
_____

### Concept Mapping

The construction of and theory behind concept mapping are discussed on pages vii–ix in the front of this Study Guide. Read those pages carefully. Then consider the concepts presented in Section 18–2 and how you would organize them into a concept map. Now look at the concept map for Chapter 18 on page 178. Notice that the concept map has been started for you. Add the key facts and concepts you feel are important for Section 18–2. When you have finished the chapter, you will have a completed concept map.

## Plantlike Protists
*(pages 394–401)*

### SECTION REVIEW

In this section you learned about the five phyla of plantlike protists. The members of phyla Euglenophyta, Pyrrophyta, and Chrysophyta are usually photosynthetic and are considered to be types of algae. The phyla Acrasiomycota and Myxomycota contain slime molds, which are not photosynthetic.

Euglenophytes are flagellates closely related to zoomastiginans. The most famous members of this group belong to the genus *Euglena*. A euglena usually swims using the longer of its two flagella. It can also creep along a surface by a process known as euglenoid movement. Although euglenas are usually photosynthetic autotrophs, they can also live as heterotrophs.

Pyrrophytes are also known as dinoflagellates. Most dinoflagellates are photosynthetic and move by means of two flagella. Many are luminescent. Dinoflagellates are the only eukaryotes that lack histones, the proteins associated with DNA.

Chrysophytes have cell walls that contain pectin instead of cellulose and store food in the form of oil rather than starch. Most chrysophytes are diatoms, protists that form intricate cell walls rich in silicon.

Acrasiomycetes are also known as cellular slime molds. They spend most of their lives as single ameboid cells. At one point in their life cycle, however, the individual ameboid cells come together to form a multicellular mass that acts much like a single organism.

Myxomycetes, or acellular slime molds, also spend part of their lives as ameboid cells. At one point in their life cycle, acellular slime molds form a multinucleate structure known as a plasmodium.

### Relating Terms: Building Vocabulary Skills

In your own words, define the terms in each of the following pairs. Then explain how the paired terms are related to each other.

**1.** Euglena, diatom: _____

_____

_____

_____

_____

_____

**2.** Dinoflagellate, bloom: _____

_____

_____

_____

_____

**3.** Acrasiomycete, myxomycete: _____

_____

_____

_____

_____

_____

_____

_____

_____

**4.** Phytoplankton, oxygen: _____

_____

_____

_____

_____

_____

_____

_____

_____

_____

**5.** Slime mold, plasmodium: _____

_____

_____

_____

_____

_____

_____

_____

_____

### Relating Form and Function: Using the Main Ideas

Examine the accompanying diagram of a euglena. In the spaces provided, identify each part of the euglena and write a definition for each part.

a. _____

_____

_____

b. _____

_____

_____

c. _____

_____

d. _____

_____

e. _____

_____

_____

_____

f. _____

_____

_____

g. _____

_____

_____

### Concept Mapping

The construction of and theory behind concept mapping are discussed on pages vii–ix in the front of this Study Guide. Read those pages carefully. Then consider the concepts presented in Section 18–3 and how you would organize them into a concept map. Now look at the concept map for Chapter 18 on page 178. Notice that the concept map has been started for you. Add the key facts and concepts you feel are important for Section 18–3. When you have finished the chapter, you will have a completed concept map.

## Using the Writing Process                                    *Chapter 18*

Use your writing skills and imagination to respond to the following writing
assignment. You will probably need an additional sheet of paper to
complete your response.

Imagine that you are the leader of Protists Protesting Pollution (PPP). You
have decided to write a letter of complaint to a number of companies that are
polluting the environment. In your letter, name some specific negative actions
that the PPP plans to take if the companies do not "clean up their act."

_____

_____

_____

_____

_____

_____

_____

_____

_____

_____

_____

_____

_____

_____

_____

_____

_____

_____

_____

_____

_____

_____

_____

_____

## Concept Mapping

The concept map below has been started for you. Add the key facts and concepts for each section of the chapter to this partial concept map. When you are done, you will have a concept map for the entire chapter.

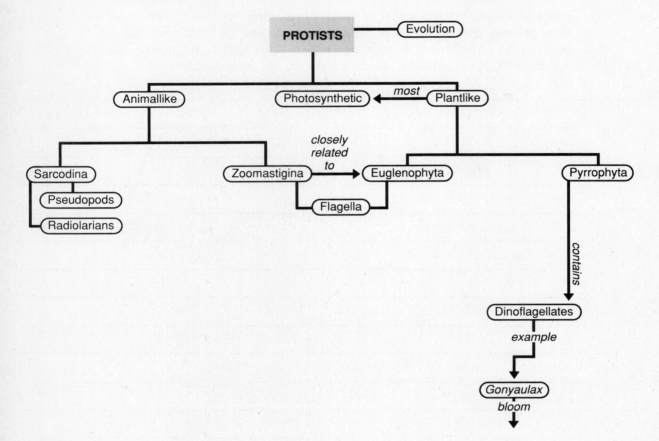

# STUDY GUIDE

| Section 19–1 | **The Fungi** | *(pages 407–416)* |

## SECTION REVIEW

In this section you learned about the interesting and sometimes exotic organisms that belong to the kingdom Fungi. You discovered that fungi are eukaryotic heterotrophs. You also discovered that many fungi are saprophytes, or organisms that obtain nourishment from decaying organic matter. Other fungi live as parasites, and still others live as symbionts.

You learned that fungi are classified into five phyla according to their methods of reproduction and their basic structure. The phylum Oomycota includes protistlike fungi such as the water mold. Common molds, such as bread mold, belong to the phylum Zygomycota. Sac fungi are members of the phylum Ascomycota, which includes yeasts and morels. Perhaps the most familiar phylum is Basidomycota, or the club fungi. This phylum includes the many varieties of mushrooms. The last phylum that you learned about was Deuteromycota, or the imperfect fungi. Included in this phylum are *Penicillium,* which is the source of the antibiotic penicillin, and many fungi that cause disease.

### Applying Definitions: Building Vocabulary Skills

Each of the terms below describes an important characteristic or a characteristic structure of fungi. Write a sentence to explain the meaning of each term.

**1.** Heterotroph: _____

_____

**2.** Decomposer: _____

_____

**3.** Mycelium: _____

_____

**4.** Hyphae: _____

_____

**5.** Gametangium: _____

_____

_____

**6.** Sporangia: _____

_____

### Five Phyla of Fungi: Understanding the Main Ideas

Each of the statements below describes a characteristic of one of the five phyla of fungi. In the blank before each statement, write the first letter if the phylum it describes: O for Oomycota, Z for Zygomycota, A for Ascomycota, B for Basidiomycota, and D for Deuteromycota.

_____ 1. Fungi in this phylum do not undergo sexual reproduction.

_____ 2. Mushrooms belong to this phylum.

_____ 3. These fungi are closely related to plantlike protists.

_____ 4. This is the largest phylum of Kingdom Fungi.

_____ 5. Common molds that grow on cheese and bread are members of this phylum.

_____ 6. This is the only group of fungi that produce motile spores.

_____ 7. Rhizoids and stolons characterize these fungi.

_____ 8. Yeasts are members of this phylum.

_____ 9. These fungi have what is probably the most elaborate life cycle of all fungi.

_____ 10. Sexual reproduction involves the formation of an ascus.

_____ 11. The reproductive structure is called a basidium.

_____ 12. *Penicillium* is a member of this phylum.

### Interpreting a Diagram: Life cycle of a Zygomycete

The diagram below shows the life cycle of black bread mold, a zygomycete. Refer to the diagram as you answer the questions.

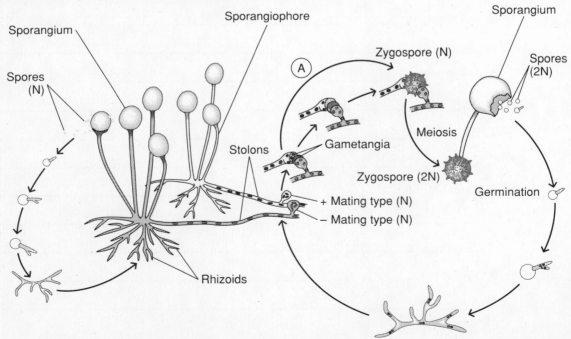

1. Three specialized types of hyphae are shown in the diagram. Name each one and describe its function. _____

_____

_____

_____

2. Which part of the diagram shows asexual reproduction? _____

_____

_____

_____

3. Explain what happens in asexual reproduction._____

_____

_____

4. Which part of the diagram shows sexual reproduction? _____

_____

_____

5. Explain what is happening in section A of the cycle. _____

_____

_____

6. What is the advantage of sexual reproduction over asexual reproduction?

_____

_____

_____

### Concept Mapping

The construction of and theory behind concept mapping are discussed on pages vii–ix in the front of this Study Guide. Read those pages carefully. Then consider the concepts presented in Section 19–1 and how you would organize them into a concept map. Now look at the concept map for Chapter 19 on page 185. Notice that the concept map has been started for you. Add the key facts and concepts you feel are important for Section 19–1. When you have finished the chapter, you will have a completed concept map.

## Fungi in Nature

*(pages 417–423)*

### SECTION REVIEW

In this section you learned about the ecological importance of fungi and the effects of fungi on human life. You learned that fungi play an important role in nature as decomposers and recyclers of organic material. This recycling is important because it prevents the loss of chemical energy and enables the return of nutrients to the soil.

You learned that fungi are found almost everywhere on Earth and in almost every kind of environment. An interesting aspect of fungi is that they have many ways in which to disperse spores in order to accomplish asexual reproduction. Another important aspect of fungi is their participation in symbiotic relationships with other organisms.

You discovered that fungi affect humans in many ways and that some of these effects are beneficial. For example, yeasts are important to people because of their role in baking and brewing. Mushrooms—those that are not poisonous—are a highly desirable food.

Effects of fungi that are not beneficial include diseases that affect people, plants, and animals. Among the most destructive plant diseases are potato blight, wheat rust, and corn smut. Among the familiar human conditions caused by fungi are athlete's foot, ringworm, and thrush.

### The Many Roles of Fungi: Interpreting the Main Ideas

Each of the statements below describes a situation in which a fungus is playing an important role. Read each statement, then describe the role of the fungus. If possible, identify the particular type of fungus involved.

1. You open the refrigerator planning to enjoy some of the raspberries you bought the day before but discover that most of the berries are covered with an unappetizing white

   fuzz. _____

   _____

2. A mass of tangled whitish threads known as reindeer moss grow in the arctic tundra, where most plants are scarce. This organism serves as food for reindeer, who might

   otherwise go hungry. _____

   _____

3. Although strep throat was at one time a life-threatening disease, this illness can now be treated effectively with an antibiotic that destroys pathogenic bacteria.

   _____

4. The odor of rotting meat comes not from a decaying carcass but from a living organism

   that looks like a lacy ball growing out of the ground. _____

   _____

   _____

**5.** A flat piece of dough left for a time in a warm place begins to expand, resulting in a mound that is three or four times as high as the original dough.

_____

_____

**6.** The roots of a plant that produces beautiful lavender orchids are covered with a

network of tiny fibers. _____

_____

_____

**7.** A few days after showering in the locker room of a public swimming pool, you discover several red, inflamed sores between your toes.

_____

_____

**8.** In France, a domestic pig is taken over the countryside to search for a delicacy that will

be sold to restaurants and gourmet shops. _____

_____

_____

**9.** Between 1845 and 1860, a third of Ireland's population starved to death because of a

shortage of potatoes, the principal food crop. _____

_____

_____

**10.** A farmer destroys all the barberry plants within the vicinity of his farm in order to

protect his wheat plants from disease and death. _____

_____

_____

### Concept Mapping

The construction of and theory behind concept mapping are discussed on pages vii–ix in the front of this Study Guide. Read those pages carefully. Then consider the concepts presented in Section 19–2 and how you would organize them into a concept map. Now look at the concept map for Chapter 19 on page 185. Notice that the concept map has been started for you. Add the key facts and concepts you feel are important for Section 19–2. When you have finished the chapter, you will have a completed concept map.

## Using the Writing Process                                      *Chapter 19*

Use your writing skills and imagination to respond to the following writing assignment. You will probably need an additional sheet of paper to complete your response.

Write a story entitled "Nature's Eccentrics." In this story, discuss some of the interesting and unusual aspects of fungi.

_____

_____

_____

_____

_____

_____

_____

_____

_____

_____

_____

_____

_____

_____

_____

_____

_____

_____

_____

_____

_____

_____

_____

_____

## Concept Mapping

The concept map below has been started for you. Add the key facts and concepts for each section of the chapter to this partial concept map. When you are done, you will have a concept map for the entire chapter.

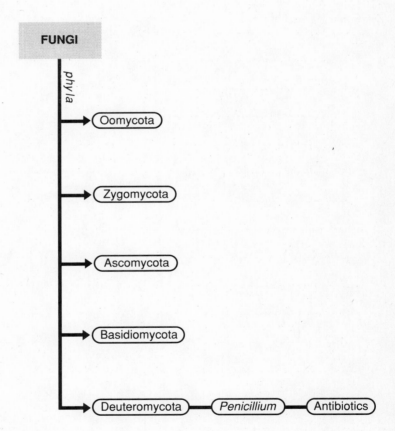

S T U D Y                                                    CHAPTER
G U I D E                                        *Multicellular Algae* **20**

**Characteristics of Algae**                        *(pages 433–435)*

### SECTION REVIEW

In this section you were introduced to algae. Algae are a diverse group of unicellular, colonial, or multicellular photosynthetic organisms that live in or near a source of water. Like land plants, algae have a cell wall and contain chlorophyll *a* as well as other pigments. Unlike most land plants, algae lack roots, stems, leaves, and an internal system of tubes to move water and materials throughout their tissues.

Algal structure, reproduction, and pigment are adapted to life under water. For example, a number of chlorophylls and accessory pigments allow algae to absorb many wavelengths of light. This enables algae to obtain energy for photosynthesis even though much light energy is absorbed by the water that surrounds them.

### Staying Alive: Finding the Main Ideas

Algae show many adaptations that enable them to survive in a watery environment. Yet on dry land these adaptations might mean that the algae could not survive. Discuss ways in which each of the following adaptations contributes to the survival of algae in water and might not contribute to the survival of algae on dry land.

**1.** Thin leaflike structures: _____

_____

_____

_____

**2.** No waterproof covering: _____

_____

_____

_____

_____

**3.** No stemlike structures: _____

_____

_____

_____

**4.** No system of internal tubes: _____

_____

_____

**5.** Accessory pigments: _____

_____

_____

_____

### Out of Step: Building Vocabulary Skills

In each of the following sets of terms, three of the terms are related. One term does not belong. Determine the characteristic common to three of the terms and then identify the term that does not belong.

**1.** Carbon dioxide, light, oxygen, wastes: _____

_____

**2.** Streams, ponds, lakes, deserts: _____

**3.** Roots, stems, cell walls, leaves: _____

### Formulating a Hypothesis

Examine the experiment shown in the accompanying illustration.

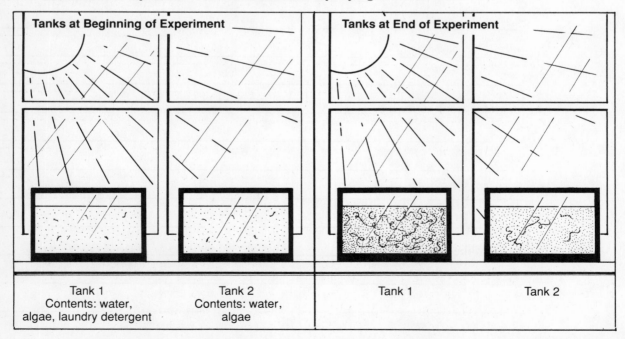

| Tanks at Beginning of Experiment | | Tanks at End of Experiment | |
|---|---|---|---|
| Tank 1 | Tank 2 | Tank 1 | Tank 2 |
| Contents: water, algae, laundry detergent | Contents: water, algae | | |

The student who performed this experiment came to the following conclusion: Plant growth was greater in the tank that contained laundry detergent than in the tank that did not.

**1.** Based on the conclusion, what was the original hypothesis in this experiment?

_____

_____

_____

**2.** What was the variable in this experiment? _____

_____

_____

**3.** What assumptions can you make about the size of the tank, the amount of sunlight shining on the tank, and the number of algae placed in the tank at the beginning of the experiment? _____

_____

_____

_____

**4.** Which of the tanks is the control? _____

_____

_____

_____

**5.** Based on this experiment, another scientist states that all plants will grow better if they are watered with soapy water. Is this reasonable? Why? _____

_____

_____

_____

_____

_____

_____

### Concept Mapping

The construction of and theory behind concept mapping are discussed on pages vii–ix in the front of this Study Guide. Read those pages carefully. Then consider the concepts presented in Section 20–1 and how you would organize them into a concept map. Now look at the concept map for Chapter 20 on page 198. Notice that the concept map has been started for you. Add the key facts and concepts you feel are important for Section 20–1. When you have finished the chapter, you will have a completed concept map.

| Section 20–2 | Groups of Algae | (pages 435–439) |

## SECTION REVIEW

In this section you learned about the three groups of multicellular algae. Algae are placed into three groups based on their color and the form in which they store food.

Green algae belong to the phylum Chlorophyta. These algae are found primarily in moist areas on land and in fresh water. All green algae contain chlorophylls *a* and *b,* store food in the form of starch, and have reproductive cycles that include sexual and asexual stages. Some species of green algae, such as *Chlamydomonas,* live as single cells. Others—*Gonium, Volvox, Spirogyra,* and *Oedogonium,* for example—are colonial. And still others, such as *Ulva,* are multicellular. Some scientists believe that the ancestors of land plants resembled living species of green algae.

Brown algae, which belong to the phylum Phaeophyta, are marine plants that are found in cool shallow coastal waters of temperate or arctic areas. Brown algae contain chlorophylls *a* and *c* and a brown accessory pigment called fucoxanthin. Brown algae store food in the form of special starches and oils. Examples of brown algae include kelp, *Sargassum,* and *Fucus.*

Red algae live in ocean waters from the far north to the tropics. These algae can live in deep water because they have special pigments that are able to trap the energy in light that penetrates ocean depths. All red algae contain chlorophyll *a* and reddish pigments called phycobilins. Some species of red algae also contain chlorophyll *d.* Most species of red algae are multicellular. All species have complicated life cycles. Examples of red algae include Irish moss (*Chondrus crispus*), coralline algae, and *Porphyra.*

### Name That Alga: Finding the Main Ideas

"It's not easy being green," laments Kermit the Frog, a popular television character. But color is everything for algae. It is one of the main characteristics scientists use to group these plants. Two characteristics are given for each of five "mystery" algae. Based on these characteristics, identify the color of each alga in the space provided.

**1.** Single cell that contains chlorophylls *a* and *b:* _____

**2.** Long plant with gas-filled bladders:_____

**3.** Plant that contains chlorophylls *a* and *c* and stores food in the form of an oil:

_____

**4.** A colony of cells that contains chlorophylls *a* and *b:* _____

**5.** Contains chlorophylls *a* and *d* and phycobilins: _____

**Separate the Terms: Building Vocabulary**

Explain the differences between the words or terms that follow. In some cases the members of the pair have very different meanings. In other cases the difference in meanings is small. Use the spaces provided for your answers.

1. Multicellular algae, colonial algae: _____

_____

_____

_____

_____

_____

2. Multicellular algae, unicellular algae: _____

_____

_____

_____

3. Green algae, red algae: _____

_____

_____

_____

_____

_____

4. Marine algae, freshwater algae: _____

_____

_____

_____

5. Chlorophyll, accessory pigments: _____

_____

_____

_____

_____

_____

_____

## How Big Is Big?: Making Calculations

Some of the larger species of brown algae can grow as much as 10 centimeters in a single day. Use your calculator to compute how long a kelp plant that grows 10 centimeters a day will be in the times shown below. Assume that the growth of this plant will not stop as long as it is alive. Begin your calculations with a plant 1 centimeter long. After each calculation, convert the length you calculated into football fields.
(*Hint:* 1 meter equals about 1.09 yards. Round off to two significant figures.)

| Time | Length in Centimeters | Length in Football Fields (100 yards) |
|---|---|---|
| One day | | |
| One week | | |
| One month (30 days) | | |
| One year (365 days) | | |
| Two years | | |
| Five years | | |

## Concept Mapping

The construction of and theory behind concept mapping are discussed on pages vii–ix in the front of this Study Guide. Read those pages carefully. Then consider the concepts presented in Section 20–2 and how you would organize them into a concept map. Now look at the concept map for Chapter 20 on page 198. Notice that the concept map has been started for you. Add the key facts and concepts you feel are important for Section 20–2. When you have finished the chapter, you will have a completed concept map.

## SECTION REVIEW

As you discovered in this section, the life cycles of most algae are much more complicated than the simple kinds of sexual reproduction that occurs in familiar animals. Algae, like other plants, typically have a life cycle that involves alternation of generations. Most algae also shift back and forth between a sexual reproduction stage that involves the production of gametes and an asexual reproduction stage that produces haploid cells called zoospores.

Asexual reproduction permits a species to colonize a new area relatively rapidly and permits the development of a population from a single individual. Sexual reproduction requires two individuals of opposite sexes to start a population; it increases the genetic variation present in a population of organisms. This increased genetic variation allows populations to adapt to changes in the environment. In algae, sexual reproduction often produces a resting stage that is capable of surviving harsh conditions that would kill a growing plant.

In the second part of this section you examined the life cycles of three different kinds of algae: the unicellular green alga *Chlamydomonas,* the multicellular green alga *Ulva,* and the multicellular brown alga *Fucus.* You also were introduced to a number of terms that describe reproduction in algae and other organisms.

### Methods of Reproduction: Finding the Main Ideas

Both sexual and asexual reproduction offer advantages and disadvantages for a species. Listed below are certain environmental situations. For each situation, decide whether it would be more advantageous for an alga to undergo sexual or asexual reproduction. Explain your answers in the space provided.

**1.** A green alga lives in a pond that is beginning to freeze. _____

_____

_____

_____

**2.** A factory that manufactures fertilizer accidentally releases plant fertilizers into a nearby lake. There is now a great deal of food for the green algae that live in the lake.

_____

_____

_____

**3.** A pond in a desert goes through alternating dry periods and wet periods during the year.

_____

_____

_____

**4.** A population of small fish with huge appetites for algae are released by mistake in a small pond. All the algae are eaten except for one plant. Suddenly all of the fish die.

_____

_____

_____

### Defining Terms: Building Vocabulary

Define the following words or phases in your own words.

**1.** Alternation of generations: _____

_____

_____

**2.** Diploid: _____

_____

_____

**3.** Haploid: _____

_____

_____

**4.** Isogamy: _____

_____

_____

**5.** Heterogamy: _____

_____

_____

_____

**6.** Sporophyte: _____

_____

_____

_____

**7.** Gametophyte: _____

_____

**8.** Egg: _____

_____

**9.** Sperm: _____

_____

**10.** Zoospore: _____

_____

### Interpreting Diagrams: Using the Main Ideas

Fill in the labels and indicate the diploid (2N) and haploid (N) stages in this diagram of the life cycle of the brown alga *Fucus*.

**Life Cycle of *Fucus***

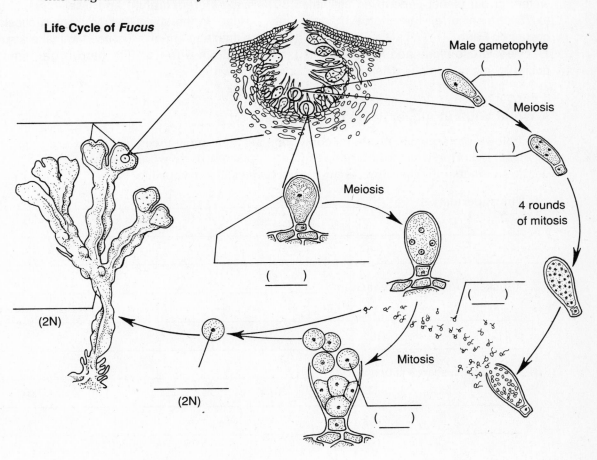

### Concept Mapping

The construction of and theory behind concept mapping are discussed on pages vii–ix in the front of this Study Guide. Read those pages carefully. Then consider the concepts presented in Section 20–3 and how you would organize them into a concept map. Now look at the concept map for Chapter 20 on page 198. Notice that the concept map has been started for you. Add the key facts and concepts you feel are important for Section 20–3. When you have finished the chapter, you will have a completed concept map.

| Section 20–4 | **Where Algae Fit into the World** | *(pages 442–443)* |

## SECTION REVIEW

Algae are a vital part of the natural world. They are a source of food for terrestrial organisms such as humans and for aquatic organisms such as fishes, sea urchins, and snails. They also provide homes for many different animals, ranging from tiny crustaceans to sea otters.

Algae produce much of the Earth's free oxygen. Oxygen is produced as a waste product of photosynthesis. The algae that live in the waters of our planet perform an estimated 50 to 75 percent of all the photosynthesis that occurs on Earth.

People use algae and the chemicals produced by algae in many different ways. Some species of algae are rich in Vitamin C, iron, iodine, and other nutrients and are used as food supplements. Other chemicals produced by algae are used to treat certain human diseases.

Algae are also used in the manufacture of many food products: ice cream, relishes, salad dressings, chip dips, and canned chow mein, to name a few. Products as diverse as toothpastes, hand lotions, finger paints, plastics, waxes, deodorants, and transistors contain algae. In the scientific laboratory, a substance extracted from algae is used to make a culture medium upon which microorganisms are grown.

### A World Without Algae: Using the Main Ideas

The following places use products containing algae or substances produced by algae. Describe what would happen in each place if there were no algae or the products of algae. How might these changes affect your life?

**1.** The supermarket: _____

_____

_____

**2.** A locker room in a gymnasium: _____

_____

_____

**3.** A medical research laboratory: _____

_____

_____

### Concept Mapping

The construction of and theory behind concept mapping are discussed on pages vii–ix in the front of this Study Guide. Read those pages carefully. Then consider the concepts presented in Section 20–4 and how you would organize them into a concept map. Now look at the concept map for Chapter 20 on page 198. Notice that the concept map has been started for you. Add the key facts and concepts you feel are important for Section 20–4. When you have finished the chapter, you will have a completed concept map.

## Using the Writing Process

Use your writing skills and imagination to respond to the following writing assignment. You will probably need an additional sheet of paper to complete your response.

> You are a reporter for a magazine called *Algae Today*. Your magazine is widely read by algae all over the world. Write an article for the magazine that would be interesting and informative to algae.

*Hint:* A few possibilities for articles include recommending a vacation resort for algae, alerting algae to certain forms of ocean pollution, giving algal beauty hints, or discussing how to deal with problem algae offspring.

_____

_____

_____

_____

_____

_____

_____

_____

_____

_____

_____

_____

_____

_____

_____

_____

_____

_____

_____

_____

## Concept Mapping

The concept map below has been started for you. Add the key facts and concepts for each section of the chapter to this partial concept map. When you are done, you will have a concept map for the entire chapter.

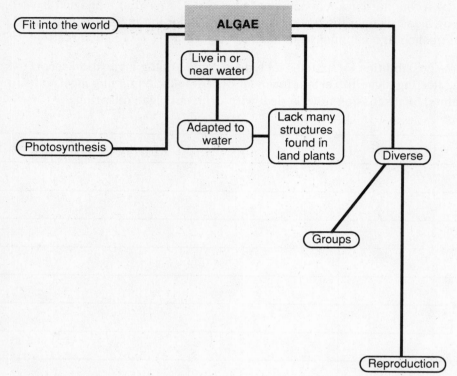

STUDY
GUIDE

| Section 21–1 | Plants Invade the Land | (pages 449–451) |

### SECTION REVIEW

In this section you read about the earliest land plants. You also learned about major challenges that plants had to meet in order to survive on land.

The evolution of land plants was a long, slow process. Algae that could live out of water appeared about 500 to 600 million years ago. From these algae ancestors emerged at least two groups of land plants. One of these groups evolved into the bryophytes: mosses, liverworts, and hornworts. Another group evolved into the tracheophytes: ferns and the rest of the higher plants.

Both bryophytes and tracheophytes have adaptations that enable them to survive on land by acquiring water and nutrients from the environment; preventing water loss; exposing the plant to the sunlight needed for photosynthesis; supporting the body of the plant; transporting water, nutrients, and other materials; allowing for gas exchange; and permitting reproduction in the absence of standing water. It is important to keep in mind that some land plants lack some of these adaptations. Not all land plants have completely solved the problems of life on land.

### Designer Plants: Finding the Main Ideas

Design a plant that is adapted to the demands of life on land. Draw your plant in the space provided. Label and briefly describe the features that enable your plant to survive on land. *Note:* You may wish to draw your plant on a separate sheet of paper. If your plant has not completely solved the problems of living on land, describe what adaptations it lacks. Explain how the lack of these adaptations restricts the kinds of environments in which your plant can live.

### Concept Mapping

The construction of and theory behind concept mapping are discussed on pages vii–ix in the front of this Study Guide. Read those pages carefully. Then consider the concepts presented in Section 21–1 and how you would organize them into a concept map. Now look at the concept map for Chapter 21 on page 208. Notice that the concept map has been started for you. Add the key facts and concepts you feel are important for Section 21–1. When you have finished the chapter, you will have a completed concept map.

## SECTION REVIEW

Bryophytes, like the algae from which they evolved, have life cycles that involve an alternation of generations between a haploid gametophyte and a diploid sporophyte. Also like the algae, these plants need water for reproduction to occur. Thus they can survive only in wet areas or in areas where a great deal of rain falls for at least part of the year. Bryophytes include mosses, liverworts, and hornworts.

Mosses, liverworts, and hornworts do not grow well outside of wet habitats for several reasons. They lack the water-conducting tubes that are found in higher plants. They lose water quickly to the surrounding air because their "leaves" lack a waterproof covering and are only one cell thick. And they also lack true roots, so they are not very efficient at taking in water from the soil.

The reproductive cycle of bryophytes is tied to a source of water because they have sperm cells that must swim through water to fertilize the eggs. In the last part of this section you examined the life cycle of a typical bryophyte: the moss *Mnium*. Recall that moss gametophytes have reproductive structures at their tips. One structure, the antheridium, produces sperm. The other structure, the archegonium, produces eggs. When a sperm joins with an egg, a diploid zygote is produced. The zygote grows into a diploid sporophyte plant, which grows on top of the gametophyte plant and is dependent on it for water and nutrients. The sporophyte produces haploid spores within a capsule. If a spore lands in a favorable place when it is released, it grows into an algalike mass of tiny filaments called a protonema. The protonema eventually develops into a gametophyte plant.

### Small Plants, Big Ideas: Finding the Main Ideas

Bryophytes are small plants that are easily overlooked. However, they are important because they were the first plants to live on land.

1. Using your own words, explain how reproduction in bryophytes is similar to that in algae. _____

_____

_____

_____

2. Describe places you would most likely find bryophytes growing. Why are these places good for bryophytes? _____

_____

_____

_____

_____

_____

3. Describe a place in which you would probably not find bryophytes growing. Explain

why bryophytes would not grow in this place. _____

_____

_____

_____

_____

_____

### Sequencing Events: Building Vocabulary Skills

Diagrams A through H show parts of the life cycle of a moss. However, they
are not in order.

*Part A.* Label the structures in Diagrams A through H.

*Part B.* Put Diagrams A through H in the correct order by writing their corresponding
letters in proper sequence in the table on the following page. In the proper place in the
table, briefly describe the structures and events shown in each diagram. The first row
has been filled out to help you get started.

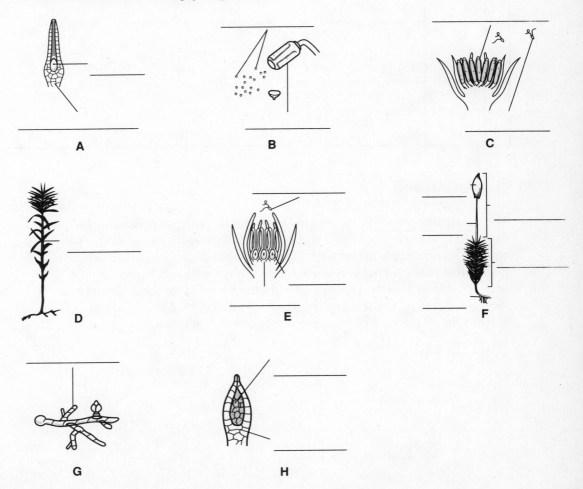

| | Diagram | Description |
|---|---|---|
| 1 | D | Mature gametophyte possesses reproductive structures at its tip. |
| 2 | | |
| 3 | | |
| 4 | | |
| 5 | | |
| 6 | | |
| 7 | | |
| 8 | | |

## Concept Mapping

The construction of and theory behind concept mapping are discussed on pages vii–ix in the front of this Study Guide. Read those pages carefully. Then consider the concepts presented in Section 21–2 and how you would organize them into a concept map. Now look at the concept map for Chapter 21 on page 208. Notice that the concept map has been started for you. Add the key facts and concepts you feel are important for Section 21–2. When you have finished the chapter, you will have a completed concept map.

**Section 21-3**   ## The Ferns and the First Vascular Plants                    *(pages 455–459)*

### SECTION REVIEW

In this section you were introduced to the first "true" land plants, the tracheophytes. Members of this phylum have evolved ways of freeing themselves from dependence on wet environments. Tracheophytes have tissues specialized for internal transport, or vascular tissues. There are two types of vascular tissues: xylem and phloem. Xylem conducts water from the roots to all parts of the plant. Phloem transports nutrients and the products of photosynthesis around the plant. Tracheophytes are also characterized by true roots, stems, and leaves.

The first vascular plants, known as psilophytes, were small and lived close to the surface of the ground. Psilophytes had primitive vascular tissues but lacked true roots and leaves. Other groups of primitive tracheophytes include lycophytes (club mosses) and sphenophytes (horsetails). Only a few species of these two groups are alive today.

Unlike lycophytes and sphenophytes, which also evolved about 400 million years ago, ferns have been quite successful. There are more than 11,000 species of ferns alive today. Ferns have true vascular tissues, strong roots, creeping or underground stems called rhizomes, and large leaves called fronds. Even though ferns can survive in drier areas than mosses, the reproduction of ferns still depends upon water for sperm to swim to eggs. The life cycles of ferns, like those of other plants, involve alternation of generations. The large plants we recognize as ferns are the diploid sporophytes. Fern gametophytes are tiny, inconspicuous heart-shaped plants.

### Standing Tall: Building Vocabulary Skills

Vascular plants such as ferns grow taller than bryophytes. In your own words, tell how each of the following structures enables vascular plants to overshadow the smaller bryophytes.

**1.** Vascular tissues: _____

_____

_____

**2.** Tracheids: _____

_____

_____

**3.** Cuticle: _____

_____

**4.** Vascular cylinder: _____

_____

**5.** Veins: _____

_____

_____

### Two Faces of Ferns: Finding the Main Ideas

**1. a.** What does a fern sporophyte look like?

_____

_____

_____

_____

**b.** Draw a sporophyte in the box labeled **Sporophyte.** Label the rhizome, roots, frond, and sori.

**2. a.** What does a fern gametophyte look like?

_____

_____

_____

**b.** Draw a gametophyte in the box labeled **Gametophyte.** Label the prothallium, rhizoids, archegonia, and antheridia.

**3.** Which generation can better survive drying? Explain.

_____

_____

_____

**4.** How does the sporophyte generation of a fern differ from the sporophyte generation of a moss?

_____

_____

_____

_____

_____

**Sporophyte**

**Gametophyte**

### Concept Mapping

The construction of and theory behind concept mapping are discussed on pages vii–ix in the front of this Study Guide. Read those pages carefully. Then consider the concepts presented in Section 21–3 and how you would organize them into a concept map. Now look at the concept map for Chapter 21 on page 208. Notice that the concept map has been started for you. Add the key facts and concepts you feel are important for Section 21–3. When you have finished the chapter, you will have a completed concept map.

**Where Mosses and Ferns Fit into the World** *(pages 460–461)*

## SECTION REVIEW

In this section you briefly reviewed how mosses and ferns fit into the natural world. You then read about a few of the ways in which these plants affect humans.

Mosses and ferns are used by gardeners in several ways. Some are grown for their beautiful shapes and colors. Others, such as sphagnum moss and peat moss, are added to garden soils to make the soil better able to support the growth of other plants.

After certain kinds of mosses die and are subjected to enormous pressure for long periods of time, they form a type of coal called peat. The energy produced by burning peat can be used to heat homes and generate electricity.

A few types of ferns are eaten directly by humans. Young fern fronds, or fiddleheads, are harvested when they appear in the early spring. These fronds can be cooked and eaten.

### Mosses and Ferns and People Too: Finding the Main Ideas

Certain newspapers have advice columns for gardeners. In these columns, an expert answers letters from readers and helps them solve their gardening problems. Suppose that you were in charge of such a column. In the space provided, respond to each of the following questions from your readers.

*Question 1:* I'd like to grow some mosses and ferns in my garden. What kinds of growing conditions are best for these plants?

*Answer:* _____

_____

_____

_____

_____

_____

*Question 2:* The dirt beneath the redwood deck in my backyard tends to wash away when it rains. Are there any kinds of ground cover plants that can be planted beneath the deck to hold the dirt in place?

*Answer:* _____

_____

_____

_____

_____

_____

**Question 3:** I usually keep my fern plant indoors, but two weeks ago I put it outside so that it could benefit from the summer sunshine. Now the edges of its fronds are turning yellow and drying up. What should I do?

**Answer:** _____

_____

_____

_____

_____

_____

_____

**Question 4:** There is an unsightly patch of moss growing beneath a large tree in my yard. How do I get rid of it? I would prefer not to use chemical pesticides.

**Answer:** _____

_____

_____

_____

_____

_____

_____

### Concept Mapping

The construction of and theory behind concept mapping are discussed on pages vii–ix in the front of this Study Guide. Read those pages carefully. Then consider the concepts presented in Section 21–4 and how you would organize them into a concept map. Now look at the concept map for Chapter 21 on page 208. Notice that the concept map has been started for you. Add the key facts and concepts you feel are important for Section 21–4. When you have finished the chapter, you will have a completed concept map.

## Using the Writing Process

Use your writing skills and imagination to respond to the following writing assignment. You will probably need an additional sheet of paper to complete your response.

Imagine that you are a moss plant growing in a forest. One day a movie agent tells you that her studio wants to make a movie of your life. But they need a diary of your experiences before they can begin. Keep a diary that includes details of your full life cycle.

_____

_____

_____

_____

_____

_____

_____

_____

_____

_____

_____

_____

_____

_____

_____

_____

_____

_____

_____

_____

## Concept Mapping *Chapter 21*

The concept map below has been started for you. Add the key facts and concepts for each section of the chapter to this partial concept map. When you are done, you will have a concept map for the entire chapter.

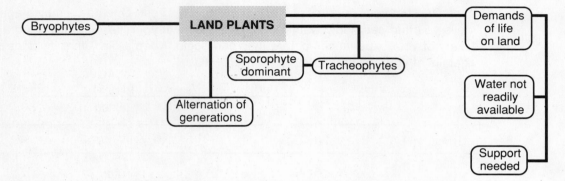

STUDY
GUIDE

| Section 22–1 | Seed Plants—The Spermopsida | (pages 467–470) |

## SECTION REVIEW

In this section you learned about the characteristics of seed plants, the spermopsida. You discovered that the members of this subphylum have numerous adaptations that allow them to survive life on land. Among these adaptations are roots, leaves, and stems; well-developed vascular tissues that conduct water and nutrients between roots and leaves; seeds; and a form of sexual reproduction that does not require standing water.

You learned that flowers and cones are the special reproductive structures of seed plants. It is within flowers and cones that the tiny gametophytes of seed plants grow and mature. The entire male gametophyte of a seed plant is contained in a structure called a pollen grain. The pollen grain is carried to the female gametophyte in a process called pollination.

In the last part of the section you learned that the zygotes of seed plants are protected by seeds. After fertilization, the zygote grows into a tiny plant called an embryo. The embryo then grows by using a supply of stored food within the seed.

### Understanding Definitions: Building Vocabulary Skills

Each of the terms listed below has a special function that contributes to the successful growth and reproduction of seed plants. Match each term in the left column with its correct function in the right column.

_____ Xylem

_____ Seed coat

_____ Pollination

_____ Phloem

_____ Flowers and cones

_____ Pollen grain

_____ Seeds

**A.** Provide nourishment for growing plant embryo

**B.** Provide place for gametophytes to grow and mature

**C.** Carries water and dissolved nutrients from the roots of plants to stems and leaves

**D.** Surrounds the plant embryo and protects it

**E.** Contains the male gametophyte

**F.** Process by which the male gametophyte is carried to the female gametophyte

**G.** Transports products of photosynthesis from one part of the plant to another

**Roots, Stems, and Leaves: Understanding the Main Ideas**

Each blank on the diagram below corresponds to one of the questions that follow. Read each question, then write the answer on the matching blank. When you are finished, you will have a labeled diagram that shows the most important adaptations of roots, stems, and leaves.

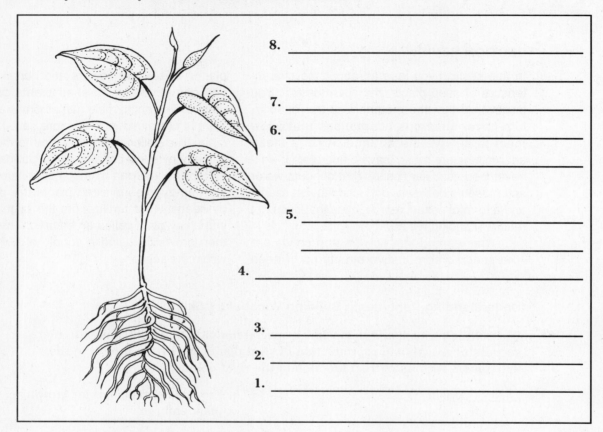

8. _____

_____

7. _____

6. _____

_____

5. _____

_____

4. _____

3. _____

2. _____

1. _____

    **1.** How do roots help nourish a plant?

**2–4.** What are three ways in which roots support and protect a plant?

    **5.** What is the main function of stems?

    **6.** Why are leaves vital to photosynthesis?

    **7.** What adaptation slows down evaporation of water from leaves?

    **8.** What substances enter and leave the leaf as needed? What adaptation allows this?

**Concept Mapping**

The construction of and theory behind concept mapping are discussed on pages vii–ix in the front of this Study Guide. Read those pages carefully. Then consider the concepts presented in Section 22–1 and how you would organize them into a concept map. Now look at the concept map for Chapter 22 on page 218. Notice that the concept map has been started for you. Add the key facts and concepts you feel are important for Section 22–1. When you have finished the chapter, you will have a completed concept map.

## Section 22–2 Evolution of Seed Plants (pages 471–475)

### SECTION REVIEW

In this section you learned how seed plants evolved. You learned that the first seed plants to appear on Earth were seed ferns. Seed ferns were soon followed by the first cone-bearing plants, or gymnosperms. The gymnosperms have grown and flourished for several hundred million years. Many of these plants are important on Earth today. Flowering plants, or angiosperms, are relative newcomers to Earth. These plants first appeared during the early Cretaceous Period, about 125 million years ago.

As you continued studying this section you learned about the major characteristics of gymnosperms and angiosperms. You discovered that gymnosperms have reproductive structures called scales, which are grouped together in cones. The seeds are not covered by the cones but sit exposed on the scales.

You learned that unlike the seeds of gymnosperms, the seeds of angiosperms are protected by a fruit. You also learned that angiosperms can be separated into two subclasses, the monocots and the dicots. The most basic difference between these two groups is that monocots have one seed leaf when they first begin to grow, whereas dicots have two seed leaves.

### Applying Definitions: Building Vocabulary Skills

Listed below are ten important terms from this section. Choose the correct term or terms from the list to answer the questions that follow.

scales              gymnosperms        pollen cones
flowers             angiosperms        fruit
monocots            cotyledons         dicots
vascular bundles

**1.** Which terms are related to reproduction in angiosperms? _____

_____

**2.** Which terms are related to reproduction in gymnosperms? _____

_____

**3.** Which term refers to the first leaf that appears when a plant embryo begins to grow?

_____

**4.** Which terms describe two subclasses of flowering plants? _____

_____

**5.** Which term is related to xylem and phloem tissues? _____

_____

**6.** Which term includes trees such as pine and spruce? _____

**7.** Which term refers to plants such as tomatoes, apple trees, and orchids?

_____

### Interpreting a Time Line: Exploring the Main Ideas

The time line below shows the appearance of various types of seed plants.
Use the time line to answer the questions that follow.

| Time (millions of years ago) | 430–395 | 395–345 | 345–280 | 280–225 | 225–190 | 190–136 | 136–125 |
|---|---|---|---|---|---|---|---|
| Era | ← Paleozoic → | | | | ← Mesozoic → | | |
| Period | Silurian | Devonian | Carboniferous | Permian | Triassic | Jurassic | Early Cretaceous |
| Plant Life | First land plants | Tree ferns grow into forests; first seed ferns appear; earliest conifers appear | Mosses and tree ferns continue to grow into forests; seed ferns continue to grow | Many cone-bearing plants appear | Cycads appear | Cone-bearing trees thrive | First flowering plants appear |

**1.** What were the first types of seed-bearing plants to appear on Earth?

_____

**2.** How long ago and during what era and period did these plants appear?

_____

**3.** What changes did tree ferns and other spore-bearing plants undergo during the

Devonian and Carboniferous periods? _____

**4.** During what period did the earliest conifers appear? _____

**5.** During what periods and how long ago did other cone-bearing plants appear?

_____

_____

**6.** How long ago and during what period did the first flowering plants appear?

_____

_____

**7.** About how long after the first plants appeared on Earth did flowering plants appear?

_____

**8.** About how many years were cone-bearing plants on Earth before flowering plants

appeared? _____

_____

### Gymnosperms and Angiosperms

Each of the statements below describes either gymnosperms or angiosperms. If a statement describes gymnosperms, place a G in the blank before the statement. If a statement describes angiosperms, place an A in the blank before the statement.

_____ 1. Include plants that are commonly called evergreens

_____ 2. Are the most widespread of all land plants

_____ 3. Palm trees are members of this group.

_____ 4. Have reproductive structures called scales

_____ 5. Some of these plants grow to more than 100 meters tall.

_____ 6. Produce male and female cones

_____ 7. Cyads are members of this group.

_____ 8. Seeds are contained within a structure called a fruit.

### Comparing Monocots and Dicots

The table below compares monocots and dicots. Fill in the missing information to complete the table.

| Monocots | Dicots |
|---|---|
| Veins in leaves _____ . | Veins in leaves _____ . |
| Flower parts are in _____ or multiples of _____ . | Flower parts are in _____ or multiples of _____ . |
| In the stem, vascular bundles are _____ _____ . | In the stem, vascular bundles are _____ _____ . |
| Seeds have _____ cotyledon(s). | Seeds have _____ cotyledon(s). |
| In the root, xylem _____ _____ ; phloem _____ _____ . | In the root, xylem _____ _____ ; phloem _____ _____ . |

### Concept Mapping

The construction of and theory behind concept mapping are discussed on pages vii–ix in the front of this Study Guide. Read those pages carefully. Then consider the concepts presented in Section 22–2 and how you would organize them into a concept map. Now look at the concept map for Chapter 22 on page 218. Notice that the concept map has been started for you. Add the key facts and concepts you feel are important for Section 22–2. When you have finished the chapter, you will have a completed concept map.

**Section 22–3** **Coevolution of Flowering Plants and Animals** *(pages 476–481)*

## SECTION REVIEW

In this section you took a second look at the evolution of seed plants. You discovered that flowering plants evolved at about the same time as the earliest mammals, after the earliest insects and birds. You learned that important relationships developed between flowering plants and a wide variety of animal species. These relationships evolved in a process that is known as coevolution. In coevolution, two organisms evolve structures and behaviors in response to changes in each other over time.

It is because of coevolution that various types of animals act as pollinators in the reproduction of flowering plants. Insects, birds, and mammals are attracted to flowering plants as a source of food. In turn, these animals carry pollen from one flower to another.

Coevolution has also made possible seed dispersal. Seed dispersal is the process by which seeds are distributed away from the parent plant. Although the seeds of some plants are dispersed by the wind, seeds of other plants are carried by animals. For example, some fruits have sharp barbs that become attached to the fur or feathers of mammals or birds. Other fruits are eaten by animals who excrete the indigestible seeds and thus deposit them in a new environment, where they can grow.

**Understanding Definitions: Building Vocabulary Skills**

1. What is vector pollination? Why is it important to flowering plants?

_____

_____

_____

_____

_____

_____

_____

2. Discuss two reasons why seed dispersal is important to plants.

_____

_____

_____

_____

_____

_____

_____

**Relating Concepts: Exploring the Main Ideas**

The chart below shows several characteristics that have evolved among flowering plants and their pollinators. Use the chart to answer the questions that follow.

| Plant Characteristic | Animal Characteristic |
|---|---|
| 1. Flower color | Bees and birds are attracted by bright colors. |
| 2. Flower odor | Moths, flies, and others have an excellent sense of smell. |
| 3. Flower size and shape | Size and shape of body parts |
| 4. "Landing platform" | Bees gather nectar only while standing. |
| 5. Nectar composition | Diet requires ingredients found in nectar. |
| 6. Nectar deep inside flower | Long tongue of moth |

**1.** What senses enable pollinating animals to find flowers? _____

_____

_____

_____

_____

**2.** What characteristics of flowers make these senses useful? _____

_____

_____

_____

_____

**3.** In what ways have flowering plants and moths evolved to benefit each other?

_____

_____

_____

_____

_____

_____

**4.** What special adaptations of flowers make them attractive to bees?

_____

_____

_____

_____

_____

**5.** How has coevolution made flowers a source of food for animals?

_____

_____

_____

_____

_____

_____

**6.** How has the physical structure of flowers coevolved with the body structures of

pollinating animals? _____

_____

_____

_____

_____

_____

### Concept Mapping

The construction of and theory behind concept mapping are discussed on pages vii–ix in the front of this Study Guide. Read those pages carefully. Then consider the concepts presented in Section 22–3 and how you would organize them into a concept map. Now look at the concept map for Chapter 22 on page 218. Notice that the concept map has been started for you. Add the key facts and concepts you feel are important for Section 22–3. When you have finished the chapter, you will have a completed concept map.

**Using the Writing Process**

Use your writing skills and imagination to respond to the following writing assignment. You will probably need an additional sheet of paper to complete your response.

A group of plants and the animals that serve as their vector pollinators have decided that they need a legal contract stating their responsibilities to one another. They have chosen you, a famous lawyer, to draw up the contract. Write a contract that assures that both sides know their responsibilities and liabilities.

_____

_____

_____

_____

_____

_____

_____

_____

_____

_____

_____

_____

_____

_____

_____

_____

_____

_____

_____

_____

_____

_____

## Concept Mapping                                                          *Chapter 22*

The concept map below has been started for you. Add the key facts and
concepts for each section of the chapter to this partial concept map. When
you are done, you will have a concept map for the entire chapter.

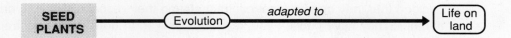

© Prentice-Hall, Inc.

S T U D Y
G U I D E

| Section 23–1 | **Soil: A Storehouse for Water and Nutrients** | *(pages 487–490)* |

### SECTION REVIEW

In this section you learned about the importance of soil for plant growth. It is from soil that plants absorb the water and nutrients that they need in order to grow.

You learned that soil is a complex mixture of sand, silt, clay, and bits of decaying animal and plant tissue. Different types of soil include sandy soil, which is composed mostly of sand grains; clay soil, which is made up of very fine clay particles; and loamy soil, which contains decaying organic matter. You also learned that soil is arranged in layers, which together make up a soil profile.

In the last part of this section you read about the essential plant nutrients found in soil. These nutrients include nitrogen, phosphorus, potassium, calcium, magnesium, and trace elements such as iron, sulfur, and zinc.

### Relating Cause and Effect: Applying the Main Ideas

Listed below are the essential plant nutrients found in soil. Each of the drawings that follow shows a plant that lacks the proper amount of an essential nutrient. Use the list and your knowledge of inorganic plant nutrients to determine the cause of each plant's illness.

nitrogen      phosphorous      potassium (potash)
calcium      magnesium      trace elements

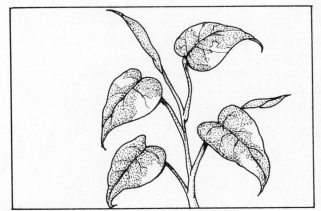

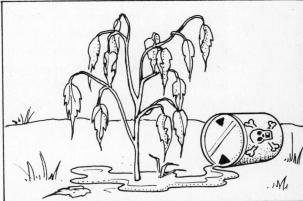

White patches on this plant's leaves indicate a lack of chlorophyll.

Cause: _____

_____

_____

This plant is dying because the soil has become toxic.

Cause: _____

_____

_____

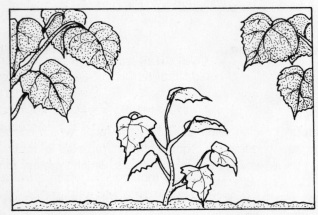

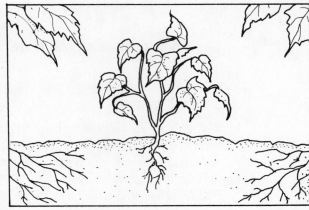

The leaves of this plant are a sickly yellow color and the plant is much too short and small.

Cause: _____

_____

_____

This plant shows stunted roots.

Cause: _____

_____

_____

## Concept Mapping

The construction of and theory behind concept mapping are discussed on pages vii–ix in the front of this Study Guide. Read those pages carefully. Then consider the concepts presented in Section 23–1 and how you would organize them into a concept map. Now look at the concept map for Chapter 23 on page 232. Notice that the concept map has been started for you. Add the key facts and concepts you feel are important for Section 23–1. When you have finished the chapter, you will have a completed concept map.

**Section 23–2** | **Specialized Tissues in Plants** | *(pages 491–493)*

## SECTION REVIEW

In this section you learned that seed plants have various types of tissue, each of which is specialized to perform a certain function. The five main types of plant tissue are meristematic, surface, parenchyma, sclerenchyma, and vascular.

Meristematic tissue enables plant roots and stems to grow in length. It is the only type of plant tissue that produces new cells by mitosis. Surface, or epidermal, tissue forms the outer surface of leaves, stems, and roots. Some epidermal tissue prevents water loss, whereas other epidermal tissue helps in the absorption of water. Parenchyma cells carry out photosynthesis and store the products of photosynthesis. Sclerenchyma cells have thick cell walls that strengthen and support a plant. Vascular tissue includes xylem and phloem. These tissues form the transport system within a plant.

### Using Analogies: Understanding the Main Ideas

In an analogy, pairs of words have a similar relationship. For example, "dark is to black as short is to small" is an analogy in which the words in each pair are similar.

Key terms from this section are listed below. Choose the term from the list that correctly completes each of the analogies that follow. You may use some terms more than once.

| | | |
|---|---|---|
| meristematic tissue | companion cells | sclerenchyma |
| cork cambium | pericycle | tracheid |
| parenchyma | apical meristem | sieve tube elements |
| vascular tissue | vascular cambium | surface (epidermal) tissue |
| vessel element | | |

**1.** Parenchyma is to storage as _____ is to support.

**2.** _____ is to growth as vascular tissue is to transport.

**3.** Tracheids are to xylem tissue as _____ are to phloem tissue.

**4.** Cambium is to thickening and branching as _____ is to length.

**5.** Parenchyma is to potatoes as _____ is to linen.

**6.** _____ is to water absorption as xylem is to water transport.

### Concept Mapping

The construction of and theory behind concept mapping are discussed on pages vii–ix in the front of this Study Guide. Read those pages carefully. Then consider the concepts presented in Section 23–2 and how you would organize them into a concept map. Now look at the concept map for Chapter 23 on page 232. Notice that the concept map has been started for you. Add the key facts and concepts you feel are important for Section 23–2. When you have finished the chapter, you will have a completed concept map.

## SECTION REVIEW

The first root that a plant puts out is called the primary root. Other roots, which branch out from the primary root, are called secondary roots. In some plants, the primary root grows longer and thicker than the secondary roots. This type of primary root is called a taproot. In other plants no single root grows longer than the others. Such plants are said to have fibrous root systems.

The tissues of mature roots can be divided into three groups: epidermis, cortex, and vascular cylinder. Root hairs in the epidermis absorb water and minerals from the soil. Water and nutrients move through the epidermis and into the cortex. From the cortex, water and nutrients move toward the center of the root into the vascular cylinder. Once water and nutrients enter the vascular cylinder, water and dissolved nutrients are forced upward into the stem of the plant.

**Completing and Interpreting a Diagram: Understanding the Main Ideas**

The diagram below shows various types of root tissue. Label each type of tissue by writing the correct term from the list on each blank. Then use the diagram to answer the questions that follow.

Casparian strip     cortex     endodermis
epidermis     phloem     root hairs
vascular cylinder     xylem

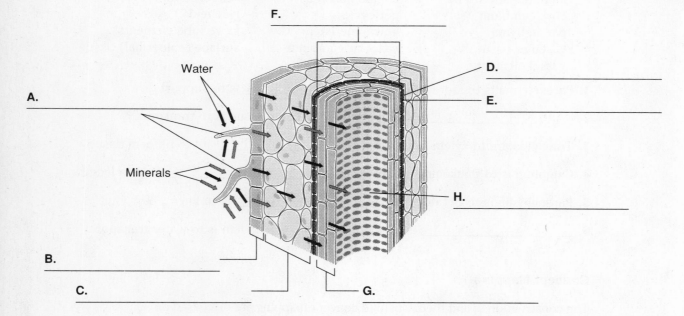

1. According to the diagram, how do water and minerals enter a root? What path do they

   take once they enter? _____

   _____

   _____

      

**2.** What happens to water and minerals once they reach their destination in the root?

_____

_____

_____

**3.** What is the purpose of the cells at point D? _____

_____

_____

**4.** What types of tissue make up the section of the root labeled at point F?

_____

_____

### Applying Definitions: Building Vocabulary Skills

Decide whether each of the following statements describes a plant with a taproot or a plant with a fibrous root system. If the statement refers to a taproot, write TR in the blank before the statement. If the statement refers to a fibrous root system, write FR in the blank.

_____ **1.** Although you pull up every dandelion in your lawn, at least half of them grow back.

_____ **2.** To prevent soil erosion, grass is planted along the side of a hill.

_____ **3.** A farmer plants rye plants between his crops to hold the topsoil in place.

_____ **4.** Roots of oak trees can reach underground water supplies far from the Earth's surface.

_____ **5.** A single plant is discovered to have 14 million secondary roots.

_____ **6.** Carrots have a short, thick root that stores sugars and starches.

_____ **7.** You grow onions in your garden and discover that they can be pulled up easily.

### Concept Mapping

The construction of and theory behind concept mapping are discussed on pages vii–ix in the front of this Study Guide. Read those pages carefully. Then consider the concepts presented in Section 23–3 and how you would organize them into a concept map. Now look at the concept map for Chapter 23 on page 232. Notice that the concept map has been started for you. Add the key facts and concepts you feel are important for Section 23–3. When you have finished the chapter, you will have a completed concept map.

| Section 23–4 | Stems | (pages 499–502) |

## SECTION REVIEW

Stems can be as small as the stem of a 2-centimeter crocus or as tall as the trunk of a 90-meter tree. Regardless of size, stems serve two main functions: They lift the leaves of plants up to the sun and they conduct various substances between roots and leaves.

Stems have four basic types of tissue: parenchyma, vascular, cambium, and cork.

Parenchyma tissue is often used for storage. Vascular tissue conducts water, nutrients, and various plant products up and down the plant. Vascular cambium produces xylem and phloem and enables a stem to grow thicker. Cork cambium produces cork, which serves as an outer protective covering for the stem.

### ▨ Interpreting a Diagram: Understanding the Main Ideas

A tree trunk is an example of a stem. The main parts of a tree trunk are *cork, sapwood, phloem, heartwood,* and *cambium.* Use these terms to label the diagram that follows. Then read the statements that follow the diagram. In the blank before each statement, write the term that it describes.

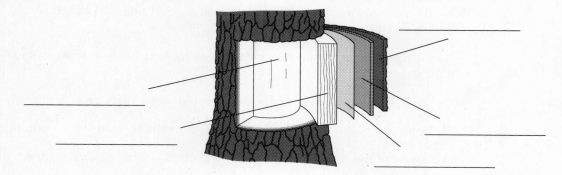

1. _____ transports the sugars produced by photosynthesis through the plant.

2. _____ enables the trunk to grow thicker by producing new xylem and phloem.

3. _____ contains active xylem tissue.

4. _____ protects the tree.

5. _____ contains xylem that no longer carries water but helps support the tree.

### ▨ Concept Mapping

The construction of and theory behind concept mapping are discussed on pages vii–ix in the front of this Study Guide. Read those pages carefully. Then consider the concepts presented in Section 23–4 and how you would organize them into a concept map. Now look at the concept map for Chapter 23 on page 232. Notice that the concept map has been started for you. Add the key facts and concepts you feel are important for Section 23–4. When you have finished the chapter, you will have a completed concept map.

| Section 23–5 | **Leaves** | *(pages 502–505)* |
| --- | --- | --- |

### SECTION REVIEW

In this section you learned that leaves are designed to collect energy from the sun and to produce food through photosynthesis. You also learned that leaves contain several specialized tissues. These include an outer covering of epidermal cells, inner layers of parenchyma cells, and vascular tissue.

Epidermal cells control water loss. They also enable the leaves to take in carbon diox-

ide and to give off oxygen. Vascular tissue in leaves is directly connected to the vascular tissue in stems. Xylem and phloem tissues in the leaves are gathered together into bundles that form the veins of a leaf. Most of the inner layers of parenchyma tissue are made up of specialized cells called leaf mesophyll. These cells contain chloroplasts and perform most of the plant's photosynthesis.

### Applying Definitions: Building Vocabulary Skills

Important terms for this section are listed below. Use these terms to label the diagram that follows.

| | | |
| --- | --- | --- |
| cuticle | guard cells | lower epidermis |
| mesophyll | palisade layer | petiole |
| spongy layer | stoma | upper epidermis |
| vein | | |

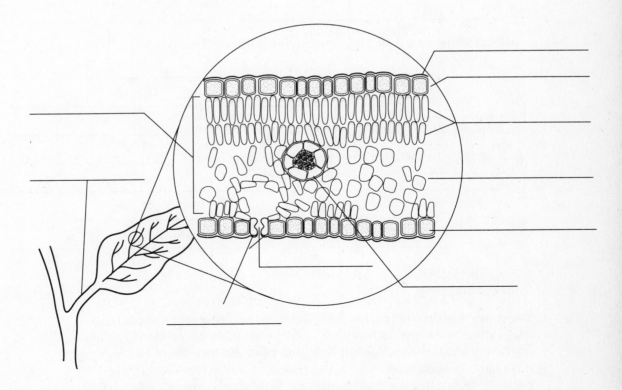

### Interpreting a Diagram: Understanding the Main Ideas

Study the diagram below, then answer the questions that follow.

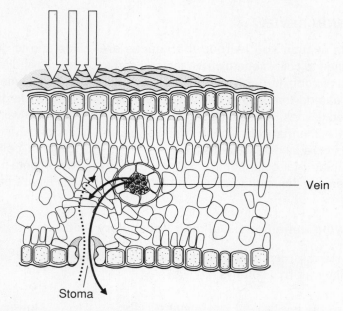

Vein

Stoma

1. Look at the arrows at the top of the diagram. What is entering the leaf at its upper

   surface? _____

2. Why is this process essential to the life of the plant? _____

   _____

3. What substance is flowing into the leaf through the stoma?

   _____

4. Where is this substance going? _____

5. What substance is leaving the leaf through the stoma? _____

6. Why is this substance shown as coming from a vein? _____

   _____

   _____

   _____

### Concept Mapping

The construction of and theory behind concept mapping are discussed on
pages vii–ix in the front of this Study Guide. Read those pages carefully. Then
consider the concepts presented in Section 23–5 and how you would organize
them into a concept map. Now look at the concept map for Chapter 23 on
page 232. Notice that the concept map has been started for you. Add the key
facts and concepts you feel are important for Section 23–5. When you have
finished the chapter, you will have a completed concept map.

## SECTION REVIEW

In this section you learned how water and dissolved nutrients are transported from the root hairs to the top of a plant. You discovered that in addition to root pressure, two other processes help to transport water and minerals. These processes are the combined action of adhesion, cohesion, and capillarity and transpiration pull. Both of these processes take place in the xylem.

You learned that it is the job of phloem to transport organic molecules and inorganic ions throughout a plant. Although the exact mechanism of phloem transport is not completely understood, it is clear that some type of active transport is involved. One hypothesis about phloem function is called the pressure flow hypothesis. It is this hypothesis that was presented in this section.

### Transpiration Pull: Interpreting the Main Idea

Examine the experiment shown below. Refer to the experiment as you answer the questions.

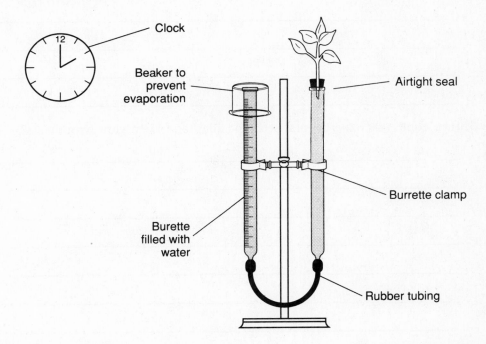

1. Based on what you have learned in this section, what do you think is being investigated in this experiment? State the problem to be investigated as a question.

_____

_____

_____

_____

**2.** After 15 minutes have elapsed, the water level in the left burette has dropped 10 milliliters. What has caused this change to occur? _____

_____

_____

_____

_____

_____

**3.** State two variables that could be introduced into this experiment to test factors that affect the problem being investigated. _____

_____

_____

**4.** State a possible hypothesis for an experiment involving one of these factors.

_____

_____

_____

_____

**5.** How could this experimental setup be adapted to test your hypothesis?

_____

_____

_____

_____

_____

_____

_____

_____

### Concept Mapping

The construction of and theory behind concept mapping are discussed on pages vii–ix in the front of this Study Guide. Read those pages carefully. Then consider the concepts presented in Section 23–6 and how you would organize them into a concept map. Now look at the concept map for Chapter 23 on page 232. Notice that the concept map has been started for you. Add the key facts and concepts you feel are important for Section 23–6. When you have finished the chapter, you will have a completed concept map.

**Section 23-7**    ## Adaptations of Plants to Different Environments    *(pages 508–511)*

### SECTION REVIEW

In this section you learned how different types of plants are adapted to survive in different environments. You learned that through natural selection, basic designs of roots, stems, and leaves have evolved in order to fulfill the particular needs of each location.

Desert plants must be adapted to dryness, heat, and sandy soil. Plants adapted to life in water must be able to grow in an environment that is nearly devoid of oxygen. Climbing plants thrive in areas such as rain forests, where plants must compete for sunlight. Some plants that live in bogs where little nitrogen is available have evolved special leaves that trap and digest insects.

### Plant Adaptations: Applying the Main Ideas

Each of the plants below is ideally adapted to one of the environments shown on the following page. Match each plant with its environment by writing the letter of the plant in the blank above its environment.

**A.** The plant has bright-red leaves that attract insects. If an insect touches hairs on the leaf, the leaf folds up suddenly, trapping the insect inside.

**B.** The plant has a root system with many root hairs that extend very deep into the soil; leaves have been reduced to thin, sharp spines.

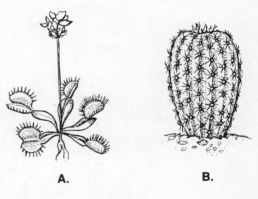

**A.**                    **B.**

**C.** The plant has aerial roots along the entire length of its stem. These roots attach to rocks or trees to help anchor the plant as it climbs.

**D.** The plant has large open spaces in long petioles that reach from the leaves down to the roots. These open spaces are filled with air through which oxygen can diffuse to the roots.

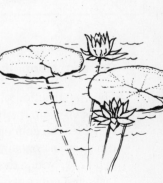

**C.**                    **D.**

1. _____

2. _____

3. _____

4. _____

## Concept Mapping

The construction of and theory behind concept mapping are discussed on pages vii–ix in the front of this Study Guide. Read those pages carefully. Then consider the concepts presented in Section 23–7 and how you would organize them into a concept map. Now look at the concept map for Chapter 23 on page 232. Notice that the concept map has been started for you. Add the key facts and concepts you feel are important for Section 23–7. When you have finished the chapter, you will have a completed concept map.

## Using the Writing Process

Use your writing skills and imagination to respond to the following writing assignment. You will probably need an additional sheet of paper to complete your response.

Imagine that you are a water molecule falling to Earth during a thunderstorm. Describe your adventures as you pass through the topsoil and enter the root of a giant redwood tree. Continue your story until you return once again to the clouds.

_____

_____

_____

_____

_____

_____

_____

_____

_____

_____

_____

_____

_____

_____

_____

_____

_____

_____

_____

_____

_____

## Concept Mapping

The concept map below has been started for you. Add the key facts and concepts for each section of the chapter to this partial concept map. When you are done, you will have a concept map for the entire chapter.

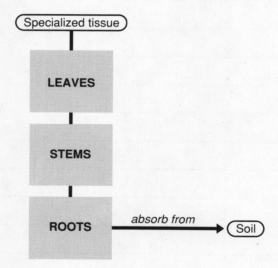

S T U D Y
G U I D E

CHAPTER
**Plant Growth and Development** 24

| Section 24–1 | **Patterns of Growth** | *(pages 517–521)* |

### SECTION REVIEW

In this section you learned that plants are classified into three main groups—annuals, biennials, and perennials—depending on how long it takes them to produce flowers and how long they live. Annuals are plants that grow from seed to maturity, then die, all in one growing season. Biennials are plants that live for two years. Perennials are plants that live for more than two years—perhaps even for thousands of years.

You also learned in this section about the growth of stems and roots. You compared the growth of monocot stems and dicot stems. You learned that roots, like stems, grow in length as their apical meristem produces new cells near the root tip. In the area just behind the root tip, the newly divided cells increase in length, pushing the root tip farther into the soil.

### Understanding Definitions: Building Vocabulary Skills

Read each of the following statements and decide if the underlined term is used correctly. If it is, write "yes" in the blank after the statement. If it is not, write the correct term to replace the underlined term.

1. Xylem that conducts water is called <u>sapwood</u>. _____

2. Tissue produced by the apical meristem is called <u>secondary phloem</u>.

   _____

3. <u>Annuals</u> live and die in one growing season.

   _____

4. The <u>root cap</u> protects the root as it forces its way through the soil.

   _____

5. <u>Biennials</u> often outlive the people who plant them.

   _____

6. Each set of <u>herbaceous tissue</u> indicates one year of a tree's growth.

   _____

7. In the <u>zone of maturation</u>, newly formed cells grow larger.

   _____

8. Xylem tissue that no longer conducts water is called <u>heartwood</u>.

   _____

▨ **Growth of Roots: Understanding the Main Ideas**

The diagram below shows a typical root tissue. Use the diagram to answer the questions.

1. In which area does most of the increase in root length occur? What is this area called?

_____

_____

_____

_____

_____

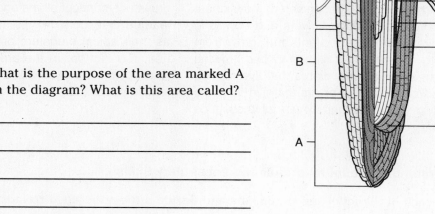

2. What is the purpose of the area marked A on the diagram? What is this area called?

_____

_____

_____

_____

_____

3. In which area do cells differentiate into mature root cells? What is this area called?

_____

_____

_____

4. At what area in the root are new cells produced? What is this area called?

_____

_____

_____

▨ **Concept Mapping**

The construction of and theory behind concept mapping are discussed on pages vii–ix in the front of this Study Guide. Read those pages carefully. Then consider the concepts presented in Section 24–1 and how you would organize them into a concept map. Now look at the concept map for Chapter 24 on page 240. Notice that the concept map has been started for you. Add the key facts and concepts you feel are important for Section 24–1. When you have finished the chapter, you will have a completed concept map.

**Section 24-2**  **Control of Growth and Development**  *(pages 521–527)*

## SECTION REVIEW

In this section you learned that division, growth, maturation, and differentiation of cells in plants is controlled by hormones. A hormone is a chemical substance produced in one part of an organism that affects another part of the organism.

Auxins are important plant hormones. They affect stem growth, root growth, and the growth of lateral buds. Other important hormones are the cytokinins, which are produced by growing roots, and gibberellins, which produce dramatic height growth in plants.

You learned that the responses of plants to environmental stimuli are called tropisms. For example, the bending of a plant toward light is called phototropism. You also learned that deciduous plants go through a variety of changes in response to the onset of winter. These changes are caused by changes in hormone production.

In the last part of this section you discovered that many plants respond to periods of light and darkness. These plants flower at different seasons of the year. Short-day plants flower when days are short, and long-day plants flower when days are long.

### Understanding Relationships: Building Vocabulary Skills

**1.** What is the relationship between a hormone and a target organ?

_____

_____

**2.** What is the relationship between auxin and lateral buds? _____

_____

**3.** What relationship exists between hormones and tropisms?_____

_____

**4.** How are bud scales and the abscission layer related? _____

_____

_____

**5.** How are phototropism and gravitropism related?_____

_____

_____

_____

### ▨ Making a Chart: Auxins

The chart that has been started below is designed to summarize several important ways in which auxins affect plant growth. Complete the chart by filling in the missing information.

| AUXINS | | | |
|---|---|---|---|
| **Target Tissue** | **Where Produced** | **Concentration** | **Action** |
| Stem | | | |
| Root | | | |
| Lateral buds | | | |
| | | | |

### ▨ Interpreting a Graph: Photoperiodicity

The graph below shows the relationship between day length and plant development in temperate regions of North America. Use the graph to answer the questions that follow.

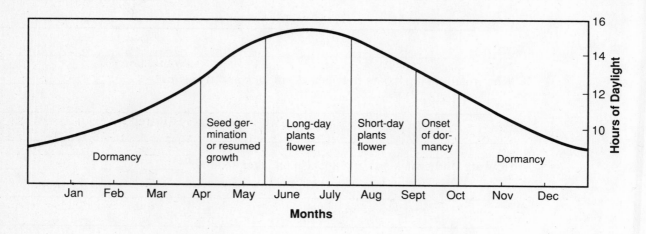

**1.** How long are days during the dormant season?_____

**2.** What is the minimum amount of daylight needed in order for plants to begin or

resume growth after dormancy?_____

**3.** How much daylight is needed in order for long-day plants to flower?

_____

**4.** During what months of the year do short-day plants flower?_____

**5.** What is the least amount of daylight needed in order for short-day plants to flower?

_____

**6.** How long must days be during the time that seeds are germinating?

_____

**7.** Express as a fraction or as a percentage the part of the year during which plants flower.

_____

### Interpreting an Experiment: Gibberellin

The drawing below shows an experiment involving four corn plants that were allowed to grow under the same conditions over a period of time. Use the drawing to answer the questions that follow.

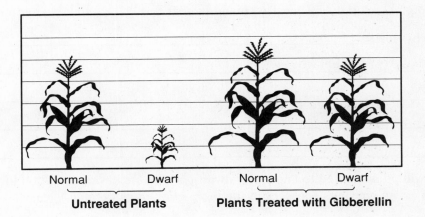

Normal          Dwarf          Normal          Dwarf
**Untreated Plants**     **Plants Treated with Gibberellin**

**1.** State in the form of a question the problem being investigated in this experiment.

_____

_____

**2.** Which plants are the control in this experiment? Which are the experimental plants?

_____

_____

_____

_____

**3.** What is the variable? _____

_____

**4.** Based on the drawing, what are your observations of the results of this experiment?

_____

_____

_____

_____

_____

**5.** Which of these observations do you find most dramatic? Why?

_____

_____

_____

_____

**6.** Based on this experiment, a scientist concludes that dwarfism in plants may be related to gibberellin deficiency. Do you agree that this is a reasonable hypothesis? Why or

why not?_____

_____

_____

_____

_____

(You will be interested to know that dwarf corn plants have a mutation for one gene that codes for an enzyme necessary in the production of gibberellin.)

## Concept Mapping

The construction of and theory behind concept mapping are discussed on pages vii–ix in the front of this Study Guide. Read those pages carefully. Then consider the concepts presented in Section 24–2 and how you would organize them into a concept map. Now look at the concept map for Chapter 24 on page 240. Notice that the concept map has been started for you. Add the key facts and concepts you feel are important for Section 24–2. When you have finished the chapter, you will have a completed concept map.

## Using the Writing Process

Use your writing skills and imagination to respond to the following writing assignment. You will probably need an additional sheet of paper to complete your response.

One day a group of plants discovered that a horticulturist had tricked them into flowering. The plants, angry over the manipulation of their natural flowering cycle, decided to sue the horticulturist. You are their attorney. State your grievances and describe how each grievance affects your client's lifestyle.

_____

_____

_____

_____

_____

_____

_____

_____

_____

_____

_____

_____

_____

_____

_____

_____

_____

_____

_____

## Concept Mapping

The concept map below has been started for you. Add the key facts and concepts for each section of the chapter to this partial concept map. When you are done, you will have a concept map for the entire chapter.

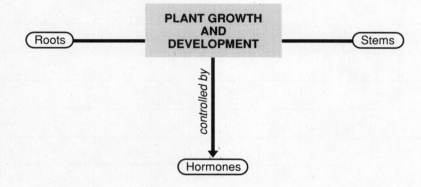

S T U D Y
G U I D E

| Section 25–1 | Cones and Flowers as Reproductive Organs | *(pages 533–540)* |

### SECTION REVIEW

The cones of gymnosperms and the flowers of angiosperms are the structures that are specialized for sexual reproduction. Mature gymnosperms produce male and female cones, which carry the structures that produce male and female gametophytes. Mature angiosperms produce flowers, and it is the flowers that contain the structures that produce gametophytes.

In this section you learned about the life cycles of gymnosperms and angiosperms. You learned how male and female gametophytes are produced and how fertilization takes place. You also learned about the importance of seeds. Seeds provide nourishment and protection for delicate embryos, and in so doing, they contribute to the survival of a plant species.

### Relating Definitions: Building Vocabulary Skills

The terms in each of the following pairs or groups are related. After each pair or group, write a sentence that describes the relationship.

**1.** Calyx, sepal: _____

_____

_____

**2.** Sepal, petal, stamen, carpel: _____

_____

_____

**3.** Petals, corolla: _____

_____

**4.** Stamen, filament, anther: _____

_____

_____

**5.** Pollination, anther, stigma: _____

_____

_____

**6.** Pistil, ovary, style, stigma: _____

_____

**7.** Ovules, ovary: _____

_____

_____

**8.** Polar nuclei, sperm nuclei, endosperm: _____

_____

_____

_____

### Making a Diagram: Using the Main Ideas

The parts of a typical flower are listed below. Use the terms in the list to label the diagram. Then use the diagram to answer the questions that follow.

| | | | |
|---|---|---|---|
| anther | filament | ovary | stigma |
| ovule | petal | pistil | style |
| pollen | sepal | stamen | |

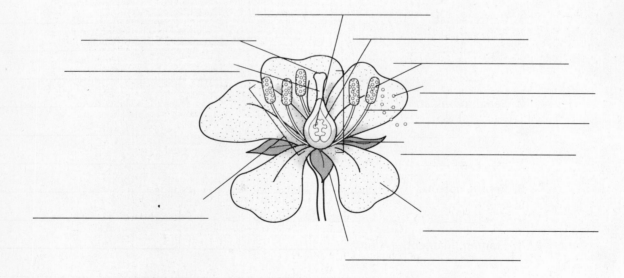

**1.** Which parts of the flower enclose the flower bud before it opens?

_____

**2.** Which parts of the flower are called sterile leaves? _____

_____

**3.** Which parts of the flower make up the fertile leaves? _____

_____

**4.** Which part of the flower contains pollen?_____

_____

**5.** On which flower part will pollen be deposited during pollination?

_____

**6.** Which parts of the flower make up the male reproductive organ?

_____

**7.** Which parts of the flower make up the female reproductive organ?

_____

### Interpreting Diagrams: Formation of Seeds

The diagram below shows the seeds of two different plants. Refer to the diagram to answer the questions that follow.

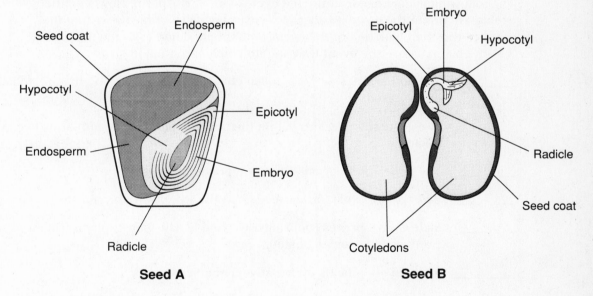

Seed A        Seed B

**1.** Which of the seeds is a monocot? Which is a dicot? How can you tell?

_____

_____

**2.** Which of the structures will develop into the plant's stem?

_____

_____

**3.** Which structure will become the primary root of the plant?

_____

_____

**4.** Why does seed A show an endosperm, whereas seed B does not? Which structure in

seed B will take over the function of the endosperm? _____

_____

_____

_____

**5.** What is the purpose of the seed coat? _____

_____

_____

_____

### Sequencing Events: Angiosperm Life Cycle

The following events occur in the life cycle of an angiosperm. However, the
events are not in the correct order. Place the events in order by writing the
number 1 in the blank before the event that should come first, the number
2 in the blank before the event that should come second, and so on.

_____ The anther dries out, the pollen chambers split open, and mature pollen grains
are released.

_____ The generative nucleus within the pollen grain divides and forms two sperm
nuclei.

_____ The pollen grain begins to grow a pollen tube.

_____ Parts of the ovule form a seed coat; the ovary wall thickens to become the fruit.

_____ Male gametophytes (pollen grains) and female gametophytes (embryo sacs) are
produced in a series of steps.

_____ A pollen grain lands on the stigma of a flower.

_____ One sperm nucleus fuses with the egg nucleus to form the zygote; the other
sperm nucleus fuses with the two polar nuclei to form the endosperm.

_____ The sperm nuclei enter the embryo sac and double fertilization occurs.

### Concept Mapping

The construction of and theory behind concept mapping are discussed on
pages vii–ix in the front of this Study Guide. Read those pages carefully. Then
consider the concepts presented in Section 25–1 and how you would organize
them into a concept map. Now look at the concept map for Chapter 25 on
page 248. Notice that the concept map has been started for you. Add the key
facts and concepts you feel are important for Section 25–1. When you have
finished the chapter, you will have a completed concept map.

| Section 25-2 | **Seed Development** | *(pages 540–542)* |

## SECTION REVIEW

In this section you learned how seeds germinate. You learned that absorbed water causes the seed coat to crack and through the seed coat the radicle emerges and grows into the primary root. You also learned about the various ways in which the seed parts of monocots and dicots push up through the soil.

You discovered that some seeds must undergo a period of dormancy before they begin to grow. Dormancy is a period during which the embryo is alive but not growing. Dormancy usually ends when environmental factors become favorable to plant growth.

### Applying Concepts: Understanding the Main Ideas

1. Using the terms listed below, make a flow chart to show what happens to a seed when it first begins to germinate.

    cotyledons         endosperm        primary root
    radicle            seed coat        water

2. List several ways in which seed dormancy contributes to the survival of a species.

_____

_____

_____

_____

_____

### Concept Mapping

The construction of and theory behind concept mapping are discussed on pages vii–ix in the front of this Study Guide. Read those pages carefully. Then consider the concepts presented in Section 25–2 and how you would organize them into a concept map. Now look at the concept map for Chapter 25 on page 248. Notice that the concept map has been started for you. Add the key facts and concepts you feel are important for Section 25–2. When you have finished the chapter, you will have a completed concept map.

| Section 25–3 | Vegetative Reproduction | (pages 542–545) |

## SECTION REVIEW

In this section you learned that in addition to reproducing sexually, many flowering plants can reproduce asexually by vegetative reproduction. Recall that vegetative reproduction does not contribute to the genetic diversity of the species. However, vegetative reproduction does enable a plant that is well-adapted to a particular environment to produce many offspring genetically identical to itself.

Vegetative reproduction occurs regularly in nature. For example, strawberry plants and bamboo plants send out long stems that pro-

duce new roots and shoots as they spread. Other plants develop tiny plants on their leaves or stems. In still other plants, new plants can develop from the detached leaves of the parent plant.

Plant growers can produce exact copies of plants with desirable characteristics through artificial vegetative reproduction. Some commonly used techniques of artificial vegetative reproduction include making cuttings, layering, and grafting.

### Vegetative Reproduction: Understanding the Main Ideas

Each of the following statements describes a type of vegetative reproduction. If the statement describes natural vegetative reproduction, write an N in the blank next to the statement. If the statement describes artificial reproduction, write a C to describe a cutting, an L to describe layering, and a G to describe grafting.

_____ 1. A stem that is cut partway through receives water and nutrients from the parent plant while it develops its own roots.

_____ 2. The kalanchoe plant produces small plants along the edges of its leaves.

_____ 3. A stem from a begonia plant is placed in water; soon it develops roots.

_____ 4. Stolons are the stems of strawberry plants that produce roots when they touch the ground.

_____ 5. A piece of stem is cut from a parent plant and attached to another plant.

### Concept Mapping

The construction of and theory behind concept mapping are discussed on pages vii–ix in the front of this Study Guide. Read those pages carefully. Then consider the concepts presented in Section 25–3 and how you would organize them into a concept map. Now look at the concept map for Chapter 25 on page 248. Notice that the concept map has been started for you. Add the key facts and concepts you feel are important for Section 25–3. When you have finished the chapter, you will have a completed concept map.

## Using the Writing Process                            *Chapter 25*

Use your writing skills and imagination to respond to the following writing assignment. You will probably need an additional sheet of paper to complete your response.

One day a group of fertile leaves decided that they were superior to all other leaves. After all, the fertile leaves argued, they were the ones responsible for reproduction. As you might expect, the sterile leaves did not like their attitude one bit. They felt they were just as important as the fertile leaves. Describe, in your own words, the debate that followed.

_____

_____

_____

_____

_____

_____

_____

_____

_____

_____

_____

_____

_____

_____

_____

_____

_____

_____

_____

_____

_____

## Concept Mapping *Chapter 25*

The concept map below has been started for you. Add the key facts and concepts for each section of the chapter to this partial concept map. When you are done, you will have a concept map for the entire chapter.

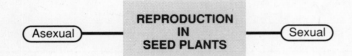

S T U D Y
G U I D E                                                       CHAPTER
                    *Sponges, Cnidarians, and Unsegmented Worms*  **26**

| Section 26–1 | **Introduction to the Animal Kingdom** | *(pages 555–560)* |

## SECTION REVIEW

With this section you began your study of the animal kingdom. You learned that animals can be classified as vertebrates or invertebrates, depending on whether or not they have a backbone. You also learned that all animals share certain basic characteristics that distinguish them from other organisms.

You discovered that an important characteristic of animals is cell specialization and division of labor. It is this division of labor that enables an animal to perform the basic functions that are essential to its survival. These functions include feeding, respiration, internal transport, elimination of waste products, response to environmental conditions, movement, and reproduction.

In the last part of this section you learned about trends in animal evolution. You discovered that some of the simplest animals exhibit radial symmetry, whereas most complex animals have bilateral symmetry. You also learned that more complex animals are characterized by a concentration of sense organs and nerve cells in their head region.

### Formulating a Definition: Building Vocabulary Skills

Use the five terms listed below to write your own definition of the word *animal*. You may use one sentence or several sentences. (Do *not* copy the definition of an animal from your textbook!)

eukaryotic          multicellular          cell walls
heterotroph          cells

_____

_____

_____

_____

_____

_____

_____

_____

_____

_____

### Relating Concepts: Understanding the Main Ideas

The seven essential life functions of an animal are listed below. Each of the statements that follow refers to one of these functions. In the blank before each statement, write the life function to which the statement refers. You may use some functions more than once.

feeding            respiration            internal transport
excretion          response               reproduction
                   movement

_____ **1.** A pumping organ called a heart forces a fluid called blood through a series of blood vessels.

_____ **2.** In some species, eggs hatch into larvae, which later undergo a process called metamorphosis.

_____ **3.** Sense organs, such as eyes and ears, gather information from the environment.

_____ **4.** Some animals are carnivores, whereas others are herbivores.

_____ **5.** Harmful wastes from cellular metabolism must be eliminated.

_____ **6.** The combination of an animal's muscles and skeleton is called its musculoskeletal system.

_____ **7.** Some species of animals bear their young alive, whereas others lay eggs.

_____ **8.** The cells of an animal must consume oxygen and give off carbon dioxide.

### Concept Mapping

The construction of and theory behind concept mapping are discussed on pages vii–ix in the front of this Study Guide. Read those pages carefully. Then consider the concepts presented in Section 26–1 and how you would organize them into a concept map. Now look at the concept map for Chapter 26 on page 258. Notice that the concept map has been started for you. Add the key facts and concepts you feel are important for Section 26–1. When you have finished the chapter, you will have a completed concept map.

| Section 26-2 | **Sponges** | *(pages 560–563)* |

## SECTION REVIEW

In this section you learned about the characteristics of sponges, which belong to the phylum Porifera. You discovered that these animals are among the most ancient on Earth and that they inhabit almost all areas of the sea.

Sponges are so different from other animals that they were once thought to be plants. They barely move, and they have no specialized tissues or organ systems and nothing that resembles a mouth or a gut. Most biologists believe that sponges evolved from single-celled ancestors separately from other multicellular animals.

Sponges are filter feeders that sift microscopic particles of food from water. The body of a sponge is designed so that water flowing through a central cavity serves as the respiratory, excretory, and internal transport systems.

### Applying Definitions: Building Vocabulary Skills

**A.** Use the terms in the accompanying list to label the diagram.

**B.** In the space provided, write the term that best matches each of the following definitions.

amebocyte
central cavity
collar cell
epidermal cell
osculum
pore
pore cell
spicule

_____ **1.** The area enclosed by the body wall of the sponge

_____ **2.** A special kind of cell that builds spicules

_____ **3.** Cells that have flagella and trap food particles.

_____ **4.** One of thousands of openings in the body wall

_____ **5.** Large hole where water leaves the sponge

_____ **6.** One of many structures that form the skeleton of the sponge

_____ **7.** Specialized cell through which water enters the sponge

_____ **8.** Cell on the outer surface of the sponge

### Form and Function: Understanding the Main Ideas

Explain in one or two sentences how sponges carry out each of the following life functions.

1. Feeding: _____

_____

_____

_____

2. Internal transport: _____

_____

_____

_____

3. Excretion: _____

_____

_____

_____

4. Respiration: _____

_____

_____

_____

5. Reproduction: _____

_____

_____

_____

_____

### Concept Mapping

The construction of and theory behind concept mapping are discussed on pages vii–ix in the front of this Study Guide. Read those pages carefully. Then consider the concepts presented in Section 26–2 and how you would organize them into a concept map. Now look at the concept map for Chapter 26 on page 258. Notice that the concept map has been started for you. Add the key facts and concepts you feel are important for Section 26–2. When you have finished the chapter, you will have a completed concept map.

**Section 26–3**    **Cnidarians**    *(pages 564–569)*

## SECTION REVIEW

In this section you were introduced to the phylum Cnidaria. You discovered that cnidarians are soft-bodied animals with stinging tentacles arranged in circles around their mouths. Some familiar cnidarians include jellyfish, corals, and hydras.

You learned that all cnidarians exhibit radial symmetry and have specialized cells and tissues. You also learned that a typical cnidarian has an internal space called a gastrovascular cavity, in which digestion takes place.

You discovered that almost all cnidarians capture and eat small animals by using stinging structures called nematocysts, which are located on their tentacles. You also learned that cnidarians lack a centralized nervous system and muscle cells. There are, however, specialized epidermal cells that serve the same function as muscle cells.

In the last part of this section, you read about the three classes of cnidarians. You also learned how cnidarians fit into the world.

### Applying Definitions: Building Vocabulary Skills

Most cnidarians have life cycles that involve two different body forms.
Label each diagram below with the name of the correct body form. Then label both diagrams to show the following parts:

|  |  |  |
|---|---|---|
| epidermis | gastroderm | gastrovascular cavity |
| mesoglea | mouth | tentacle |

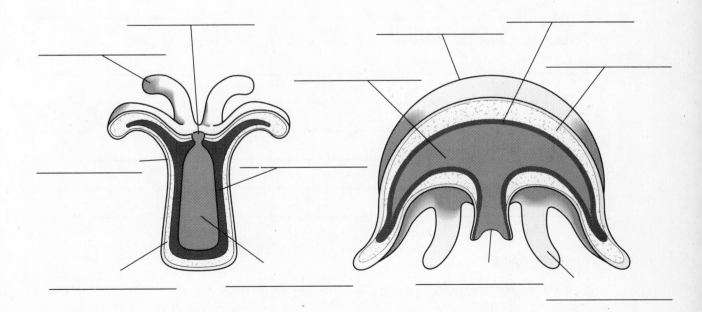

**Interpreting Diagrams: Exploring the Main Ideas**

Use the accompanying diagrams to answer the questions that follow.

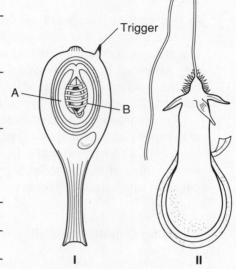

1. Where on the body of a cnidarian are these

    structures located? _____

2. What occupies the region labeled A on the

    diagram? _____

3. What is the structure labeled B?

    _____

4. Briefly describe the condition of the stinging
    cell in Figure I.

    _____

    _____

    _____

5. What is the function of the trigger?_____

    _____

    _____

    _____

    _____

6. What is the condition of the nematocyst in Figure II? What has happened?

    _____

    _____

    _____

    _____

    _____

    _____

    _____

**Concept Mapping**

The construction of and theory behind concept mapping are discussed on
pages vii–ix in the front of this Study Guide. Read those pages carefully. Then
consider the concepts presented in Section 26–3 and how you would organize
them into a concept map. Now look at the concept map for Chapter 26 on
page 258. Notice that the concept map has been started for you. Add the key
facts and concepts you feel are important for Section 26–3. When you have
finished the chapter, you will have a completed concept map.

**Section 26—4**  **Unsegmented Worms**  *(pages 570–579)*

## SECTION REVIEW

In this section you were introduced to the group of animals known as unsegmented worms. Unsegmented worms include flatworms (phylum Platyhelminthes) and roundworms (phylum Nematoda).

You learned that flatworms are the simplest animals with bilateral symmetry. You also learned that most members of this phylum exhibit enough cephalization to have what can be called a head.

You discovered that roundworms are among the simplest animals that have a digestive system with two openings, a mouth and an anus. Several parasitic roundworms that cause diseases in humans were discussed, including *Ascaris*, *Trichinella*, and hookworms.

### Understanding Definitions: Building Vocabulary Skills

Each of the statements below describes either flatworms, roundworms, or both. If the statement describes flatworms, write an F in the blank before the statement. If the statement describes roundworms, write an R. If the statement describes both, write both an F and an R.

_____ **1.** Are invertebrates

_____ **2.** Are members of phylum Nematoda

_____ **3.** Includes blood flukes

_____ **4.** Includes free-living and parasitic animals

_____ **5.** Have a digestive system with only one opening

_____ **6.** May have asexual reproduction

_____ **7.** Eliminate undigested wastes through the anus

_____ **8.** Includes *Ascaris*

### Applying Concepts: Understanding the Main Ideas

The body plan of a free-living flatworm is shown at right.

**1.** Label each lettered structure on the diagram.

**2.** Label the anterior and posterior ends of the worm.

**3.** What type of symmetry does the body show?

_____

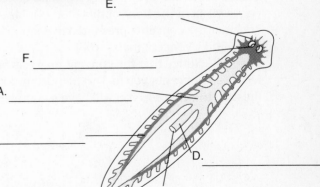

E. _____

F. _____

A. _____

B. _____

C. _____

D. _____

4. What is the purpose of the branches on structure A?_____

_____

_____

_____

5. What evidence does this diagram show of cephalization?_____

_____

_____

_____

_____

6. What is the function of the structure labeled D? _____

_____

_____

_____

_____

7. What is the function of the structure labeled F? _____

_____

_____

_____

_____

## Concept Mapping

The construction of and theory behind concept mapping are discussed on
pages vii–ix in the front of this Study Guide. Read those pages carefully. Then
consider the concepts presented in Section 26–4 and how you would organize
them into a concept map. Now look at the concept map for Chapter 26 on
page 258. Notice that the concept map has been started for you. Add the key
facts and concepts you feel are important for Section 26–4. When you have
finished the chapter, you will have a completed concept map.

## Using the Writing Process

Use your writing skills and imagination to respond to the following writing assignment. You will probably need an additional sheet of paper to complete your response.

An environmental group has suggested that sponges must stop releasing their waste products into the ocean. They want to pass a law that would require sponges to find an alternative way of disposing of their waste products.

**a.** Write the law as it might be drafted by the environmental group.

**or**

**b.** Write an article to support the right of sponges to continue their lifestyle.

_____

_____

_____

_____

_____

_____

_____

_____

_____

_____

_____

_____

_____

_____

_____

_____

_____

_____

_____

## Concept Mapping

The concept map below has been started for you. Add the key facts and concepts for each section of the chapter to this partial concept map. When you are done, you will have a concept map for the entire chapter.

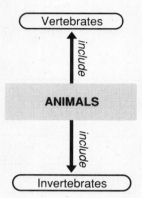

STUDY
GUIDE

CHAPTER
**27**
*Mollusks and Annelids*

| Section 27–1 | **Mollusks** | *(pages 585–593)* |

## SECTION REVIEW

What is a mollusk? In the first part of this section you discovered that a wide diversity of animals are classified as mollusks. These animals are classified together in one phylum because all show similar features during early development and all exhibit different forms of the same basic body plan.

You then explored the wide variety of form and function among mollusks. You also learned about three classes of mollusks. Gastropods, such as snails and slugs, move by means of a broad ventral foot; many have a one-piece shell. Bivalves, such as clams, have two shells that are joined by a hinge. Cephalopods, such as octopi and nautiluses, have tentacles.

Mollusks affect humans in a variety of ways. Many mollusks are popular as food, and the oyster is important not only as a food source but as a producer of pearls. Some mollusks also impact negatively on humans by serving as intermediate hosts for parasites and by doing damage to gardens and crops.

### Relating Definitions: Building Vocabulary Skills

Many of the important terms in this section relate to the way mollusks carry out basic life functions. Listed below are the seven basic functions that animals must carry out in order to survive. Following is a list of terms from this section. In the blank following each term, write the function or functions that the term relates to. Some functions may be used more than once, and others may not be used.

feeding            reproduction        response              movement
excretion         respiration          internal transport

**1.** Trochophore: _____

**2.** Foot: _____

**3.** Mantle: _____

**4.** Radula: _____

**5.** Gills: _____

**6.** Open circulatory system:

_____

**7.** Nephridia: _____

**8.** Closed circulatory system:

_____

### Classifying Mollusks: Understanding the Main Ideas

Each of the following statements describes one of the three main classes of mollusks. In the blank before each statement, write a G if the statement describes a gastropod, a B if the statement describes a bivalve, and a C if the statement describes a cephalopod.

_____ **1.** Scallops are members of this group.

_____ **2.** They feed using a structure called a radula.

_____ **3.** Sometimes they produce pearls.

_____ **4.** The chambered nautilus is a member of this group.

_____ **5.** They are highly intelligent and may be more intelligent than some vertebrates.

_____ **6.** They have two shells.

_____ **7.** Members of this group use tentacles to capture their prey.

_____ **8.** Most members of this group have small internal shells or no shells at all.

_____ **9.** Most have a broad, muscular foot located on their stomach.

_____ **10.** Most are sessile, but some can move around rapidly by flapping their shells.

### Making a Diagram: Internal Structure of a Bivalve

A clam is a typical bivalve. The diagram below shows the internal structure of a clam. Use the terms listed to correctly label the diagram.

| | | |
|---|---|---|
| Anterior adductor muscle | Intestine | Stomach |
| Gill | Ganglion | Nephridium |
| Anus | Excurrent siphon | Incurrent siphon |
| Heart | Posterior adductor muscle | Mouth |
| Foot | Gonad | |

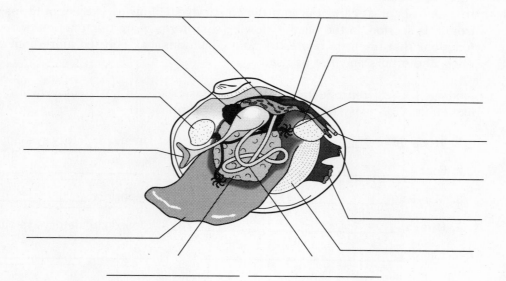

### Concept Mapping

The construction of and theory behind concept mapping are discussed on pages vii–ix in the front of this Study Guide. Read those pages carefully. Then consider the concepts presented in Section 27–1 and how you would organize them into a concept map. Now look at the concept map for Chapter 27 on page 264. Notice that the concept map has been started for you. Add the key facts and concepts you feel are important for Section 27–1. When you have finished the chapter, you will have a completed concept map.

Section 27–2   **Annelids**   *(pages 594–601)*

## SECTION REVIEW

In this section you were introduced to members of the phylum Annelida. These animals, which are also known as segmented worms, include the familiar earthworm as well as about 9000 other species, such as sandworms, bloodworms, and leeches.

You learned that annelids are characterized by a long, segmented body and that they live both in water and on land. By studying in detail the earthworm's body systems, you learned how annelids carry out essential life functions.

Annelids are important in many habitats. Small annelids that live in the ocean serve as food for other organisms. Earthworms and similar annelids are important in soil conditioning. Earthworms also perform the valuable function of processing nutrients from dead organisms into substances that can be used by plants.

**Applying Concepts: Basic Functions in Annelids**

Complete each sentence below to describe how the indicated function is carried out by annelids. You may add additional sentences if you wish.

1. Respiration: Aquatic annelids typically breathe _____

   _____

2. Internal transport: The circulatory system in annelids _____

   _____

   _____

   _____

3. Excretion: Annelids produce two kinds of wastes. Solid wastes _____

   _____

   _____

   _____

4. Response: Annelids have a well-developed nervous system _____

   _____

   _____

   _____

5. Movement: Muscles in the annelid _____

   _____

6. Reproduction: Most annelids reproduce _____

   _____

### Relating Concepts: Understanding the Main Ideas

Listed in the left column are the major organs that make up the digestive system of the earthworm. Listed in the right column are words that describe the basic function of each organ. Match each function in the right column with the corresponding organ in the left column by writing the correct letter in the blank.

_____ **1.** Intestine     **a.** Chops food into small pieces

_____ **2.** Anus     **b.** Pumps food and soil or grabs prey

_____ **3.** Gizzard     **c.** Entrance for food

_____ **4.** Crop     **d.** Eliminates wastes

_____ **5.** Esophagus     **e.** Storage area for food

_____ **6.** Pharynx     **f.** Digests food

_____ **7.** Mouth     **g.** Passageway for food

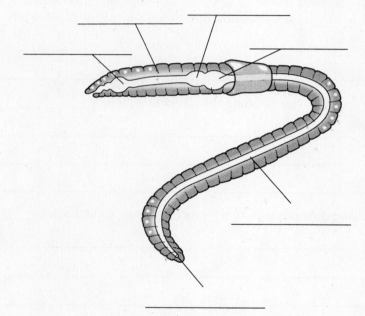

### Concept Mapping

The construction of and theory behind concept mapping are discussed on pages vii–ix in the front of this Study Guide. Read those pages carefully. Then consider the concepts presented in Section 27–2 and how you would organize them into a concept map. Now look at the concept map for Chapter 27 on page 264. Notice that the concept map has been started for you. Add the key facts and concepts you feel are important for Section 27–2. When you have finished the chapter, you will have a completed concept map.

## Using the Writing Process

Use your writing skills and imagination to respond to the following writing assignment. You will probably need an additional sheet of paper to complete your response.

A civil war is brewing in Mollusca. The gastropods and bivalves have begun a campaign that will result in all cephalopods being declared illegal aliens. The gastropods and bivalves assert that to be a legal resident of Mollusca an organism must not have tentacles.

Your assignment is to develop a public relations campaign that conclusively shows that cephalopods are legal and valuable residents of Mollusca.

_____

_____

_____

_____

_____

_____

_____

_____

_____

_____

_____

_____

_____

_____

_____

_____

_____

_____

_____

## Concept Mapping                                            *Chapter 27*

The concept map below has been started for you. Add the key facts and
concepts for each section of the chapter to this partial concept map. When
you are done, you will have a concept map for the entire chapter.

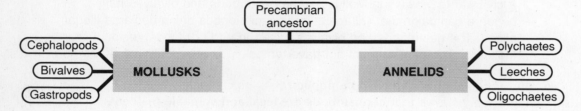

S T U D Y
G U I D E

| Section 28–1 | **Introduction to Arthropods** | *(pages 607–616)* |

### SECTION REVIEW

In this section you were introduced to the largest and most diverse animal phylum: the arthropods. You discovered that this group of animals includes horseshoe crabs, spiders, scorpions, ticks, shrimp, barnacles, waterfleas, centipedes, millipedes, and butterflies and all other insects, as well as the large group of extinct animals known as trilobites.

You learned that the three main characteristics of an arthropod are a tough exoskeleton, a series of jointed appendages, and a segmented body. You learned how arthropods carry out essential life functions, and you studied in detail the body plan of a representative arthropod, the grasshopper.

You discovered that growth and development in arthropods is associated with the process called molting. Arthropods molt, or shed, their exoskeletons in order to grow larger. Some arthropods also change in form as well as in size when they molt. Some arthropods, such as grasshoppers and crabs, change gradually as they grow and develop. Others, such as butterflies and beetles, change abruptly and dramatically as they go from one stage of development to the next.

### Relating Terms: Building Vocabulary Skills

Consider the following terms: *exoskeleton, chitin,* and *molt.* In the space provided, write a brief paragraph in which you discuss how these terms are related to one another and explain the importance of these terms in identifying arthropods.

_____

_____

_____

_____

_____

_____

_____

_____

_____

_____

### Interpreting Diagrams: Understanding the Main Ideas

The accompanying diagram shows the internal structures of a representative arthropod, the grasshopper. Refer to the diagram to answer the questions that follow.

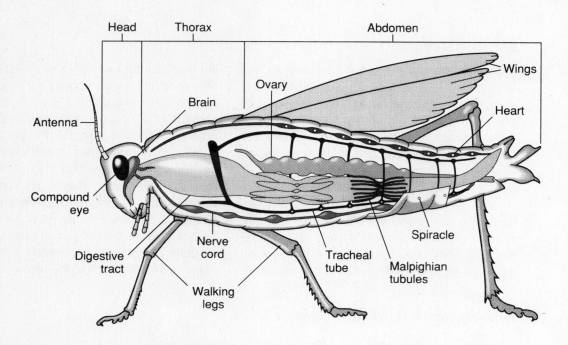

1. Which structures are found in the head of a grasshopper?_____

   _____

2. What structure connects the brain to the rest of the body?_____

   _____

3. What is the long structure that runs along the top of the abdomen? What is its function?

   _____

4. What relationship exists between a spiracle and a tracheal tube? What is the function of

   the tracheal tubes? _____

   _____

   _____

5. What is the function of the Malpighian tubules? How is their function related to their

   location?_____

   _____

   _____

**Metamorphosis: Interpreting Diagrams**

The accompanying diagrams show two types of metamorphosis that takes place in insects. Refer to the diagrams to answer the questions that follow.

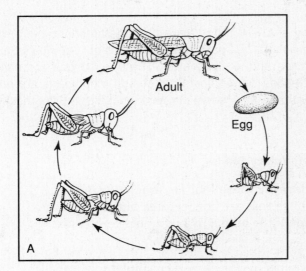

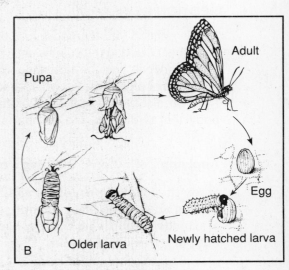

1. In diagram A, what changes do you see in the arthropod as it changes from a young animal to an adult? _____

2. Are there other changes in the animal in diagram A that you cannot see?

_____

3. Does diagram A show complete or incomplete metamorphosis? Explain.

_____

_____

4. How many different forms of growth are shown in diagram B? What are these forms?

_____

5. Does diagram B show complete or incomplete metamorphosis? Explain.

_____

_____

**Concept Mapping**

The construction of and theory behind concept mapping are discussed on pages vii–ix in the front of this Study Guide. Read those pages carefully. Then consider the concepts presented in Section 28–1 and how you would organize them into a concept map. Now look at the concept map for Chapter 28 on page 278. Notice that the concept map has been started for you. Add the key facts and concepts you feel are important for Section 28–1. When you have finished the chapter, you will have a completed concept map.

**Section 28–2**   **Spiders and Their Relatives**                              *(pages 617–620)*

## SECTION REVIEW

In this section you learned about the arthropods that belong to the subphylum Chelicerata. These animals include horseshoe crabs and the group of organisms known as arachnids.

You discovered that chelicerates have a body that is divided into two parts and that they have two characteristic pairs of mouthparts. The first pair of mouthparts are called chelicerae. The second pair are called pedipalps. Both sets of mouthparts are adapted to serve different functions in different species.

### Applying Definitions: Building Vocabulary Skills

Study the following list of animals. Identify those that are arachnids by putting an A in the blank. Identify those that are chelicerates but not arachnids by putting a C in the blank. If the animal is not a chelicerate, do not put a mark in the blank.

_____ **1.** Horseshoe crab            _____ **7.** Trilobite

_____ **2.** Grasshopper              _____ **8.** Scorpion

_____ **3.** Red velvet mite          _____ **9.** Tick

_____ **4.** Wolf spider              _____ **10.** Centipede

_____ **5.** Praying mantis           _____ **11.** Hummingbird

_____ **6.** Tarantula               _____ **12.** Chigger

### Interpreting a Diagram: Understanding the Main Ideas

The internal structures of a typical spider are shown in the accompanying diagram. Use the diagram to answer the questions that follow.

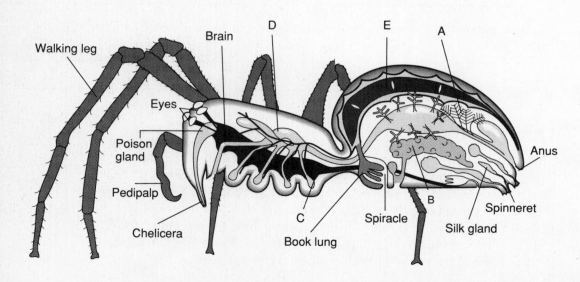

© Prentice-Hall, Inc.

1. How many major body parts does a spider have? What are they called? Label these body parts on the diagram. _____

_____

_____

_____

2. Based on the diagram, what is the relationship between the poison gland and the chelicera? What would you expect the function of the chelicera to be?

_____

_____

_____

_____

_____

_____

3. Identify structure A and describe its function. _____

_____

_____

_____

4. Which lettered structure is the heart? _____

5. What appears to be the relationship between the silk gland and the spinneret?

_____

_____

_____

_____

_____

### Concept Mapping

The construction of and theory behind concept mapping are discussed on pages vii–ix in the front of this Study Guide. Read those pages carefully. Then consider the concepts presented in Section 28–2 and how you would organize them into a concept map. Now look at the concept map for Chapter 28 on page 278. Notice that the concept map has been started for you. Add the key facts and concepts you feel are important for Section 28–2. When you have finished the chapter, you will have a completed concept map.

## Section 28–3  Crustaceans

### SECTION REVIEW

In this section you learned about the subphylum Crustacea. Crustaceans are primarily aquatic. They range in size from microscopic water fleas to spider crabs up to 6 meters across. Crustaceans are characterized by a hard exoskeleton, two pairs of antennae, and mouthparts called mandibles. The main body parts are the head, thorax, and abdomen. In many species, the head and thorax have fused into a cephalothorax that is covered by a shell called the carapace.

**Summarizing Information: Finding the Main Ideas**

**A.** Identify and label the following structures on the accompanying diagram of a crayfish: *abdomen, brain, carapace, cephalothorax, claw, first antenna, gills, heart, intestine, mandible, nerve cord, second antenna, swimmeret, tail, walking leg.*

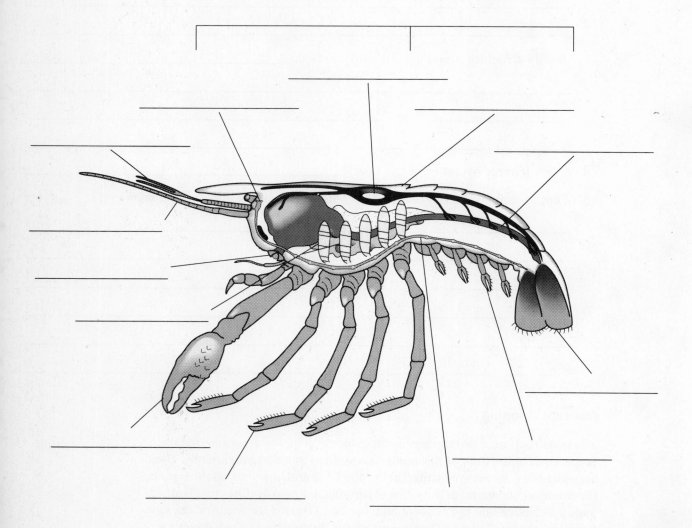

**B.** Complete the table as follows: (1) List each type of major appendage on a crayfish; (2) tell whether each type of appendage is attached to the head, thorax, or abdomen; (3) briefly describe the function of each type of appendage.

| Major Appendages on a Crayfish | | |
|---|---|---|
| **Appendage** | **Location** | **Function** |
| | | |
| | | |
| | | |
| | | |
| | | |
| | | |
| | | |

### Concept Mapping

The construction of and theory behind concept mapping are discussed on pages vii–ix in the front of this Study Guide. Read those pages carefully. Then consider the concepts presented in Section 28–3 and how you would organize them into a concept map. Now look at the concept map for Chapter 28 on page 278. Notice that the concept map has been started for you. Add the key facts and concepts you feel are important for Section 28–3. When you have finished the chapter, you will have a completed concept map.

## SECTION REVIEW

The subphylum Uniramia is made up of insects and their relatives, the millipedes and centipedes. Uniramians are arthropods that are characterized by one pair of antennae and appendages that do not branch. Uniramians inhabit almost every terrestrial habitat on Earth. In addition, some species live in fresh water and a few others live in marine habitats.

The first class of uniramians that you read about in this section, centipedes, consists of carnivores that have one pair of legs on each of their many body segments. Centipedes possess poison claws in their head region, which are used to capture and stun or kill prey.

The second class of uniramians, millipedes, consists of herbivores that have two pairs of legs on each of their many body segments. Many millipedes curl up into a ball to protect themselves. Some can also defend themselves by secreting unpleasant or toxic chemicals.

The third class of uniramians, insects, consists of arthropods that are characterized by a body composed of three parts—head, thorax, and abdomen—and that have three pairs of legs attached to the thorax. Insects have developed many intriguing adaptations for feeding, movement, social behavior, and communication.

### Describing Insects: Using the Main Ideas

Some of the following statements correctly describe insects. Others do not. Read each statement carefully. If the statement correctly describes insects, write "correct" in the space provided. If the statement does not correctly describe insects, write "incorrect" and explain why the statement is incorrect in the space provided.

**1.** The bodies of insects are characterized by two main sections and five pairs of legs.

_____

_____

**2.** Many insects undergo a developmental process called metamorphosis, which can be

incomplete or complete. _____

_____

**3.** Unlike many arthropods, insects have no mouthparts. _____

_____

_____

**4.** Many insects form societies in which members are dependent upon one another for

survival. _____

_____

**5.** Many insects use chemicals called pheromones to communicate with one other.

_____

**6.** Insects are characterized by a long, wormlike body composed of many leg-bearing

segments._____

_____

_____

_____

**7.** Almost all insects are aquatic. _____

_____

**8.** Insects have two pairs of antennae and often have a hard exoskeleton that contains

calcium carbonate. _____

_____

_____

**9.** Insects may communicate through "dancing." _____

_____

### Insect Body Plan: Labeling Diagrams

On the accompanying diagram of a typical insect, label the following structures:
*abdomen, antenna, compound eye, head, leg, mandibles, thorax, wings.*

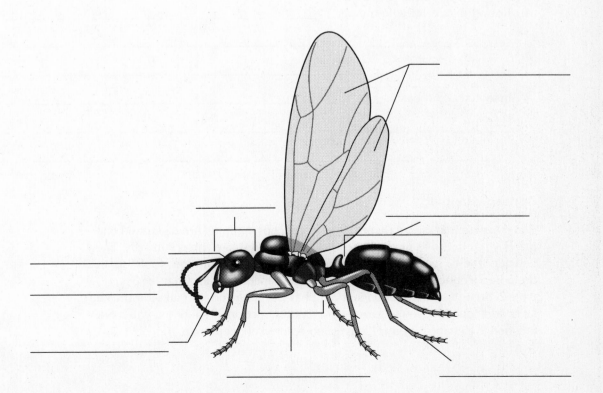

### Relating Concepts: Finding the Main Ideas

Explain how the terms in each of the following pairs are related to each other.

1. Centipedes, millipedes: _____

_____

_____

2. Uniramian, insect: _____

_____

3. Mandibles, mouthparts: _____

_____

_____

4. Queen, worker: _____

_____

_____

5. Pheromone, cricket chirp: _____

_____

_____

6. Round dance, waggle dance: _____

_____

_____

7. Insect society, caste: _____

_____

_____

### Concept Mapping

The construction of and theory behind concept mapping are discussed on pages vii–ix in the front of this Study Guide. Read those pages carefully. Then consider the concepts presented in Section 28–4 and how you would organize them into a concept map. Now look at the concept map for Chapter 28 on page 278. Notice that the concept map has been started for you. Add the key facts and concepts you feel are important for Section 28–4. When you have finished the chapter, you will have a completed concept map.

| Section 28–5 | How Arthropods Fit into the World | (pages 629–631) |
|---|---|---|

## SECTION REVIEW

In this section you learned about the many roles that arthropods play in nature. For example, they are direct and indirect sources of food for many organisms. Arthropods are also involved in many symbiotic relationships with plants and animals, including the pollination of flowers.

You discovered that arthropods affect humans both positively and negatively. Although some are important as sources of food and as agents of pollination in agriculture, others cause crop damage and carry diseases.

**Applying Concepts: Understanding the Main Ideas**

1. Describe how each of the following organisms would probably view an arthropod:

Venus fly trap: _____

_____

_____

_____

_____

Frog: _____

_____

_____

_____

_____

Flowering plant: _____

_____

_____

_____

_____

Fish: _____

_____

_____

_____

_____

Acacia tree: _____

_____

_____

_____

_____

**2.** List four ways in which arthropods can have a positive effect on humans and four ways in which they can have a negative effect.

Positive: _____

_____

_____

_____

_____

_____

_____

Negative: _____

_____

_____

_____

_____

_____

_____

### Concept Mapping

The construction of and theory behind concept mapping are discussed on pages vii–ix in the front of this Study Guide. Read those pages carefully. Then consider the concepts presented in Section 28–5 and how you would organize them into a concept map. Now look at the concept map for Chapter 28 on page 278. Notice that the concept map has been started for you. Add the key facts and concepts you feel are important for Section 28–5. When you have finished the chapter, you will have a completed concept map.

## Using the Writing Process

Use your writing skills and imagination to respond to the following writing assignment. You will probably need an additional sheet of paper to complete your response.

Many newspapers and magazines print "personals"—brief classified advertisements about a personal matter such as friendship or romance. A personal usually contains a brief description of the advertiser as well as the kind of person the advertiser wishes to meet. Advertisers usually try to be as interesting and clever as possible in order to entice people to respond to their ad.

Imagine that you are a lovelorn arthropod. Write a personal in which you describe the mate of your dreams.

*Hint:* Pretend to be a specific kind of arthropod. For example, you could decide to be a black widow spider, a deer tick, a desert scorpion, or a horseshoe crab—not just "a chelicerate." You might wish to consult your textbook, an encyclopedia, or some other reference for information about the arthropod you have selected.

_____

_____

_____

_____

_____

_____

_____

_____

_____

_____

_____

_____

_____

_____

_____

_____

_____

## Concept Mapping

The concept map below has been started for you. Add the key facts and concepts for each section of the chapter to this partial concept map. When you are done, you will have a concept map for the entire chapter.

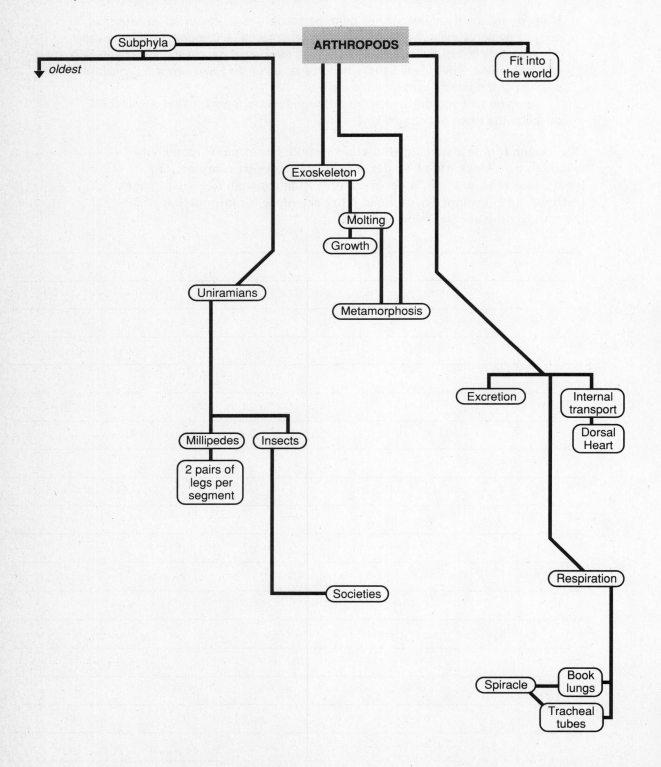

S T U D Y
G U I D E

| Section 29–1 | **Echinoderms** | *(pages 637–644)* |
|---|---|---|

## SECTION REVIEW

In this section you learned about the spiny-skinned animals that belong to the phylum Echinodermata. Echinoderms include animals such as starfishes, brittle stars, sand dollars, sea cucumbers, and sea lilies.

You discovered that echinoderms are characterized by spiny skin, five-part radial symmetry, an internal skeleton, and a unique body system called the water vascular system. The water vascular system consists of a sys-tem of internal water-filled canals and many suction-cuplike structures called tube feet. It is involved in movement and many other essential life functions.

You learned that echinoderms are important in controlling the populations of other organisms in many marine habitats. In recent years, echinoderms have become important to humans as subjects of scientific research and as possible sources of medicine.

### Formulating Definitions: Building Vocabulary Skills

1. In your own words, write a definition of the word *echinoderm* that lists five important characteristics of echinoderms.

_____

_____

_____

_____

_____

_____

2. List six animals that are classified as echinoderms.

_____

_____

_____

_____

_____

▨ **Identifying Structures: Building Vocabulary Skills**

Examine the accompanying illustration of a starfish. In the spaces provided, identify and write a brief definition for each part of the starfish. Note that the last two structures have been identified for you.

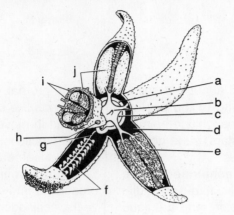

**a.** _____

_____

**b.** _____

_____

**c.** _____

_____

**d.** _____

_____

**e.** _____

_____

**f.** _____

_____

**g.** _____

_____

**h.** _____

_____

**i.** Skeletal plates: _____

_____

**j.** Gonad: _____

_____

### Characteristics of Echinoderms: Using the Main Ideas

Decide if each of the following statements correctly describes echinoderms. If it does, write an E in the blank before the statement. If it does not, write an N and explain why the statement is incorrect.

_____ **1.** Echinoderms live in salt water and in fresh water. _____

_____

_____ **2.** All echinoderms are carnivores. _____

_____

_____ **3.** Echinoderms have no brain. _____

_____

_____ **4.** Echinoderms have a respiratory system that includes lungs or feathery gills.

_____

_____ **5.** Echinoderms have an external skeleton. _____

_____

_____ **6.** Echinoderms are vertebrates. _____

_____

_____ **7.** Adult echinoderms exhibit bilateral symmetry. _____

_____

_____ **8.** The water vascular system is unique to echinoderms. _____

_____

_____ **9.** In most species of echinoderms, solid wastes are expelled through the anus.

_____

_____ **10.** Although primitive, the nervous system of the echinoderm can tell the animal when it is in light or darkness and when it is right side up.

_____

### Concept Mapping

The construction of and theory behind concept mapping are discussed on pages vii–ix in the front of this Study Guide. Read those pages carefully. Then consider the concepts presented in Section 29–1 and how you would organize them into a concept map. Now look at the concept map for Chapter 29 on page 284. Notice that the concept map has been started for you. Add the key facts and concepts you feel are important for Section 29–1. When you have finished the chapter, you will have a completed concept map.

| Section 29–2 | Invertebrate Chordates | (pages 645–646) |
|---|---|---|

## SECTION REVIEW

In this section you were introduced to the chordates, or members of the phylum Chordata. You also learned about the invertebrate chordates.

Chordates are animals that are characterized by a notochord, a hollow dorsal nerve cord, and pharyngeal slits. Some chordates possess these features as adults. Others may possess these features only at an early stage of development.

The invertebrate chordates are divided into two groups: tunicates and lancelets. Invertebrate chordates are of interest to evolutionary biologists because they represent the link between vertebrates and the rest of the animal kingdom. It is important to remember that vertebrates did not evolve from modern tunicates and lancelets. However, modern invertebrate chordates are thought to be similar to the ancient chordates that gave rise to the vertebrates.

### Defining Terms: Building Vocabulary Skills

In your own words, define each of the following terms.

1. Lancelet:_____

_____

2. Hollow dorsal nerve cord: _____

_____

3. Chordate: _____

_____

4. Tunicate: _____

_____

5. Notochord:_____

_____

6. Pharyngeal slits:_____

_____

_____

### Concept Mapping

The construction of and theory behind concept mapping are discussed on pages vii–ix in the front of this Study Guide. Read those pages carefully. Then consider the concepts presented in Section 29–2 and how you would organize them into a concept map. Now look at the concept map for Chapter 29 on page 284. Notice that the concept map has been started for you. Add the key facts and concepts you feel are important for Section 29–2. When you have finished the chapter, you will have a completed concept map.

**Using the Writing Process**                                    *Chapter 29*

Use your writing skills and imagination to respond to the following writing assignment. You will probably need an additional sheet of paper to complete your response.

You are probably familiar with the advice columns found in many newspapers and magazines. These columns consist of brief letters from readers and responses from the columnist.

Imagine that you are a friend of an echinoderm of your choice. Much as you like this echinoderm, you find some of its personal habits rather upsetting. In desperation, you decide to seek the counsel of a famous advice columnist.

Your assignment is to write both the letter to the columnist and the columnist's reply.

_____

_____

_____

_____

_____

_____

_____

_____

_____

_____

_____

_____

_____

_____

_____

_____

## Concept Mapping

The concept map below has been started for you. Add the key facts and concepts for each section of the chapter to this partial concept map. When you are done, you will have a concept map for the entire chapter.

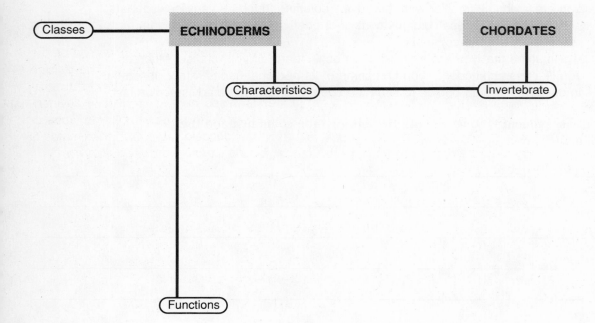

STUDY
GUIDE

| Section 30–1 | **Evolution of the Invertebrates** | *(pages 653–657)* |
|---|---|---|

## SECTION REVIEW

In this section you learned that the evolution-ary relationships among different groups of organisms can be shown in the form of a dia-gram known as a phylogenetic tree. Recall that scientists determine evolutionary relation-ships by examining fossils and comparing the embryos, body structures, and chemical com-pounds of living organisms.

There are several major branches on the phylogenetic tree of the animals. The division of animals into protostomes and deutero-stomes is based on events in early develop-ment. The division of animals into acoelo-mates, pseudocoelomates, and coelomates is based on the structure of the body cavity.

### Identifying Word Parts: Building Vocabulary Skills

Many scientific terms are made up of word parts that are derived from Latin or Greek words. Each of the following word parts forms one or more of the key terms in this section. In the space provided, write the meaning of each word part. Then give an example of a term from Section 30–1 that contains that word part.

**1.** pseudo-: _____

**2.** a-: _____

**3.** proto-: _____

**4.** deutero-: _____

**5.** coelom: _____

**6.** -stome: _____

**7.** phylo-: _____

**8.** -geny: _____

**9.** -derm: _____

**10.** meso-: _____

### Acoelomates, Pseudocoelomates, and Coelomates: Using the Main Ideas

**1.** How do acoelomates, pseudocoelomates, and coelomates differ from one another?

_____

_____

_____

**2.** In the space provided, draw a diagram that shows how acoelomates, pseudocoelomates, and coelomates differ in basic body structure. Label your diagram.

## Protostomes and Deuterostomes: Interpreting Diagrams

Fill in the blanks in the accompanying diagram. Then answer the questions on the following page.

Development of a _____

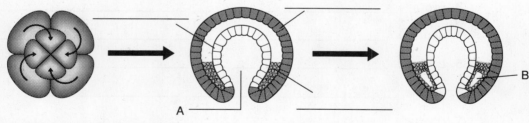

_____

Development of a _____

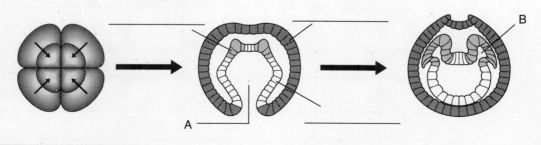

_____

1. What is structure A? _____

_____

2. What does structure A become in a protostome? _____

_____

3. What does structure A become in a deuterostome? _____

_____

4. What is structure B? _____

_____

5. How does early development differ in protostomes and deuterostomes?

_____

_____

_____

_____

_____

_____

_____

_____

_____

_____

_____

## Concept Mapping

The construction of and theory behind concept mapping are discussed on pages vii–ix in the front of this Study Guide. Read those pages carefully. Then consider the concepts presented in Section 30–1 and how you would organize them into a concept map. Now look at the concept map for Chapter 30 on page 293. Notice that the concept map has been started for you. Add the key facts and concepts you feel are important for Section 30–1. When you have finished the chapter, you will have a completed concept map.

| Section 30-2 | **Form and Function in Invertebrates** | *(pages 658-668)* |

## SECTION REVIEW

As you began studying this section you learned an important concept: Evolution is random and undirected. Organisms are not better or worse, or more perfect or less perfect, than one another. They are simply different.

In this section you compared the body systems that carry out the essential life functions in invertebrates. Some systems are simple; others are quite complex. None are "better" in any absolute sense than others. They can all be considered successful because they accomplish the functions they have evolved to perform. As you examine the functions of movement, feeding, internal transport, respiration, excretion, response, reproduction, and development, you discovered that the ways in which invertebrates carry out these functions vary greatly. You also learned that it is useful to think of the digestive, excretory, and nervous systems in invertebrates as showing an evolutionary trend toward increasing complexity and specialization.

### Recognizing Prefixes: Building Vocabulary Skills

Many of the key terms in this section contain prefixes that appear frequently in words pertaining to biology. For each term or group of terms listed below, define the underlined prefix(es) Then define the term(s).

1. Hydrostatic skeleton: _____

_____

_____

2. Endoskeleton: _____

_____

_____

3. Intracellular digestion; internal fertilization: _____

_____

_____

_____

_____

4. Extracellular digestion; external fertilization: _____

_____

_____

_____

_____

### Skeletal Systems: Exploring the Main Ideas

In the space provided, identify each type of skeleton shown in the accompanying diagram. Then answer the questions that follow.

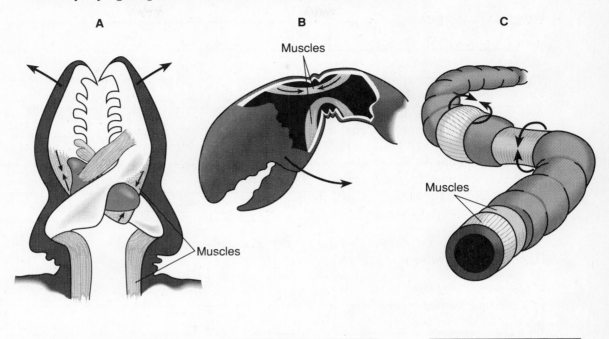

A            B            C

_____   _____   _____

1. What are the two kinds of invertebrates that have the type of skeleton shown in diagram A?

_____

_____

In diagram B? _____

_____

_____

In diagram C? _____

_____

_____

2. What is the relationship between the muscles and the supporting structures in diagram A?

_____

_____

In diagram B?_____

_____

In diagram C?_____

_____

**Relating Concepts: Understanding the Main Ideas**

1. Compare open and closed circulatory systems. _____

   _____

   _____

   _____

   _____

   Name two invertebrates that have an open circulatory system. _____

   _____

   _____

   Name two invertebrates that have a closed circulatory system.

   _____

   _____

   _____

2. Fill in the accompanying diagram by listing and briefly describing four types of respiratory structures found in invertebrates. For each type of structure, name an organism that possesses it.

| Structure | Description | Example |
|-----------|-------------|---------|
|           |             |         |
|           |             |         |
|           |             |         |
|           |             |         |

**3.** Name and briefly describe four types of excretory structures that are found in

invertebrates. _____

_____

_____

_____

_____

_____

_____

_____

_____

_____

**4.** How do the numbers of gametes and offspring produced by an invertebrate relate to its

methods of fertilization and parental care? _____

_____

_____

_____

_____

_____

_____

_____

_____

### Concept Mapping

The construction of and theory behind concept mapping are discussed on
pages vii–ix in the front of this Study Guide. Read those pages carefully. Then
consider the concepts presented in Section 30–2 and how you would organize
them into a concept map. Now look at the concept map for Chapter 30 on
page 293. Notice that the concept map has been started for you. Add the key
facts and concepts you feel are important for Section 30–2. When you have
finished the chapter, you will have a completed concept map.

## Using the Writing Process

Use your writing skills and imagination to respond to the following
writing assignment. You will probably need an additional sheet of paper
to complete your response.

When looking at the phylogenetic tree of the animals, it is easy to get the
mistaken impression that there were fewer types of animals in the remote
past. After all, there are very few branches at the trunk of the tree!
However, although there may be more species of animals alive today, there
were actually many more phyla of animals in the past. It is not clear why
some phyla have survived to the present whereas others became extinct
long ago. Many extinct animals appear to have been as fit to survive as the
animals alive today.

The accompanying illustrations show several invertebrates that belong
to phyla that are now extinct. Write a story that takes place in an alternative
Earth where these animal phyla did not die out.

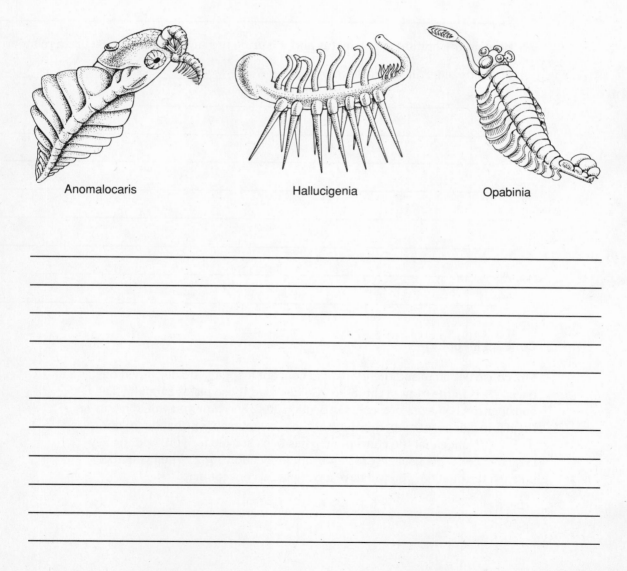

Anomalocaris          Hallucigenia          Opabinia

_____

_____

_____

_____

_____

_____

_____

_____

_____

_____

## Concept Mapping

The concept map below has been started for you. Add the key facts and concepts for each section of the chapter to this partial concept map. When you are done, you will have a concept map for the entire chapter.

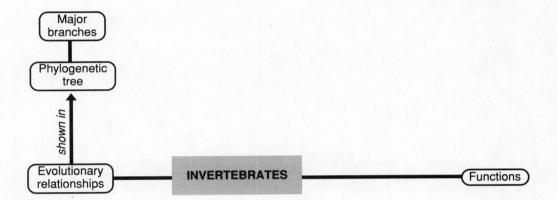

S T U D Y
G U I D E

CHAPTER

31

*Fishes and Amphibians*

| Section 31–1 | **Fishes** | *(pages 679–692)* |

### SECTION REVIEW

In this section you discovered that at some time during their development vertebrates have a notochord, a hollow dorsal nerve cord, and pharyngeal slits. In most vertebrates, the notochord is replaced by a vertebral column during development.

You learned that fishes are aquatic vertebrates that usually have scales, paired fins, and gills. Fishes have a two-chambered heart and a single-loop circulatory system. Fishes may be oviparous (egg-laying), ovoviviparous (eggs are incubated inside the mother's body), or viviparous (embryos develop inside the mother and are nourished directly by the mother's body).

Jawless fishes, such as lampreys and hagfishes, are eellike parasites and scavengers that lack paired fins, scales, and a vertebral column. They are the most primitive type of fishes.

Members of the class Chondrichthyes have a skeleton of cartilage. Cartilaginous fishes include sharks, rays, skates, sawfish, and chimaeras.

Bony fishes make up the class Osteichthyes and about 40 percent of all vertebrates. Most bony fishes are ray-finned fishes. Bony fishes include groupers, salmons, eels, and mudskippers, to name a few.

### Identifying Internal Structures: Building Vocabulary Skills

The drawing below represents the internal anatomy of a fish. Label each part of the fish.

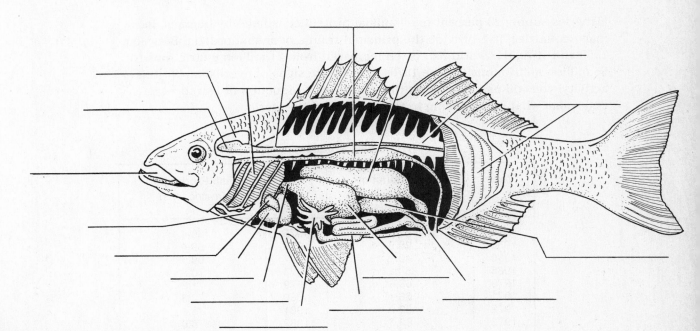

### Relating Form to Function: Finding the Main Ideas

The chart below lists some of the internal structures of the fish. In the space provided, identify the main function of each structure.

| Structure | Function |
|---|---|
| Anus | |
| Brain | |
| Esophagus | |
| Gills | |
| Heart | |
| Intestine | |
| Kidney | |
| Mouth | |
| Ovary | |
| Pyloric ceca | |
| Stomach | |
| Swim bladder | |

### Analyzing Data

Fisheries supply 23 percent of all animal protein consumed by humans. In many countries, fish provide the principal source of animal protein. Between 1950 and 1970, the world fish catch increased from 21 million metric tons to 66 million metric tons. By the 1970s, there were signs of overfishing. In this activity, you will use data about the worldwide production of fish to determine what is happening to the fisheries of the world.

**World Fish Production, 1950 – 1983**

| Year | Fish Production (metric tons) | Year | Fish Production (metric tons) |
|---|---|---|---|
| 1950 | 21.1 | 1975 | 66.4 |
| 1955 | 28.9 | 1976 | 69.4 |
| 1960 | 40.2 | 1977 | 68.5 |
| 1965 | 53.2 | 1978 | 70.2 |
| 1970 | 65.6 | 1979 | 71.2 |
| 1971 | 66.1 | 1980 | 72.3 |
| 1972 | 62.0 | 1981 | 75.1 |
| 1973 | 62.7 | 1982 | 76.8 |
| 1974 | 66.5 | 1983 | 74.0 |

**1.** How did fish production change between 1950 and 1960? _____

_____

**2.** How did fish production change between 1960 and 1970? _____

_____

**3.** How did fish production change between 1970 and 1980? _____

_____

**4.** How was the growth rate in fish production between 1950 and 1960 different from the growth rate in fish production between 1970 and 1980? Explain your answer.

_____

_____

_____

**5.** When did the fish population decrease? _____

### Recognizing Fact and Opinion

Read the following statements about fishes. Determine whether each statement is fact or fiction. Indicate your answer by the word "fact" or the word "fiction" on the line provided.

_____ **1.** A hagfish has six hearts.

_____ **2.** Some fishes are able to glow in the dark.

_____ **3.** Most sharks attack and kill humans.

_____ **4.** A shark uses about 20,000 teeth throughout its lifetime.

_____ **5.** A shark's skeleton is made up of bone..

_____ **6.** Fishes have a keen sense of smell.

_____ **7.** Electric eels can produce electric potentials of several hundred volts.

_____ **8.** All fishes use gills to get their oxygen supply.

_____ **9.** Shark skin feels like sandpaper.

_____ **10.** The whale shark is the largest fish in existence, measuring over 18 meters in length.

### Find the Oddball: Identifying Patterns

Examine each of the following lists of animals. In each list, one of the animals does not belong. In the space provided, write the name of the animal that does not belong and explain your choice.

**1.** shark, jellyfish, guppy, ray _____

_____

**2.** sawfish, shark, salmon, ray _____

_____

**3.** lamprey, trout, hagfish _____

_____

**4.** skate, eel, guppy, grouper _____

_____

### Identifying Internal Structures: Building Vocabulary Skills

The drawing below represents the brain of a fish. Label each part of the fish brain.

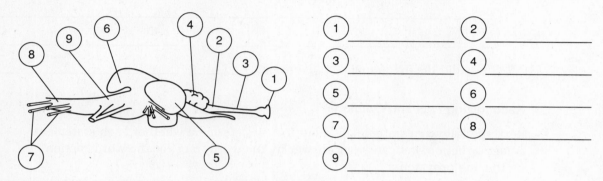

1 _____  2 _____

3 _____  4 _____

5 _____  6 _____

7 _____  8 _____

9 _____

### Relating Form to Function: Finding the Main Ideas

The chart below lists the parts of the fish brain. In the space provided, identify the main function of each structure.

| Structure | Function |
|---|---|
| Cerebellum | |
| Cerebrum | |
| Medulla | |
| Olfactory lobe | |
| Optic lobe | |

### Concept Mapping

The construction of and theory behind concept mapping are discussed on pages vii–ix the front of this Study Guide. Read those pages carefully. Then consider the concepts presented in Section 31–1 and how you would organize them into a concept map. Now look at the concept map for Chapter 31 on page 304. Notice that the concept map has been started for you. Add the key facts and concepts you feel are important for Section 31–1. When you have finished the chapter, you will have a completed concept map.

**Section 31–2**  **Amphibians**  *(pages 692–701)*

## SECTION REVIEW

In this section you learned that amphibians are vertebrates that have a moist skin, lack scales and claws, and are usually aquatic as larvae and terrestrial as adults. Most adult amphibians breathe with lungs.

Although amphibian larvae are herbivores and filter feeders, amphibian adults are usually carnivores. Adult amphibians usually use a combination of lungs, skin, and mouth cavities to breathe, whereas amphibian larvae usually use their gills and skin.

Adult amphibians have a three-chambered heart and a double-loop circulatory system. Fertilization in amphibians may be internal or external. Amphibians may be oviparous, ovoviviparous, or viviparous. Amphibians have well-developed nervous and sensory systems.

Salamanders have a tail even after they have undergone metamorphosis into the adult form. Frogs and toads have large hind legs adapted for jumping and swimming. Adult frogs and toads usually lack tails.

### Identifying Internal Structures: Building Vocabulary Skills

The drawing below represents the internal anatomy of a frog. Label each part of the frog.

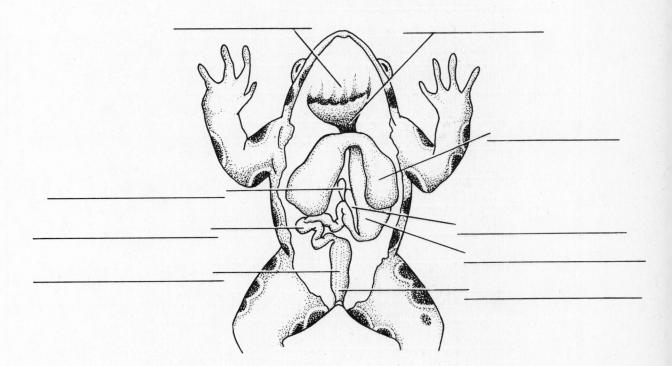

### Relating Form to Function: Finding the Main Ideas

The chart below lists some of the internal structures of the frog. In the space provided, identify the main function of each structure.

| Structure | Function |
|-----------|----------|
| Cloaca | |
| Esophagus | |
| Gall bladder | |
| Large intestine | |
| Liver | |
| Mouth | |
| Pancreas | |
| Small intestine | |
| Stomach | |

### Frog Metamorphosis: Sequencing an Event

Describe what is happening during each stage in the metamorphosis of a frog.

_____

_____

_____

_____

_____

_____

_____

_____

_____

_____

_____

_____

_____

▓ **Identifying Internal Structures: Building Vocabulary Skills**

The drawing below represents the brain of a frog. Label each part of the frog brain.

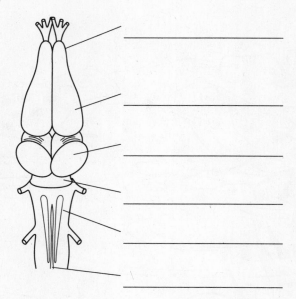

_____

_____

_____

_____

_____

_____

▓ **Comparing Body Forms**

Complete the following chart.

| Characteristic | Tadpole | Adult Frog |
|---|---|---|
| Habitat | | |
| Method of breathing | | |
| Method of locomotion | | |
| Diet | | |
| Type of circulatory system | | |

▓ **Concept Mapping**

The construction of and theory behind concept mapping are discussed on pages vii–ix in the front of this Study Guide. Read those pages carefully. Then consider the concepts presented in Section 31–2 and how you would organize them into a concept map. Now look at the concept map for Chapter 31 on page 304. Notice that the concept map has been started for you. Add the key facts and concepts you feel are important for Section 31–2. When you have finished the chapter, you will have a completed concept map.

**Using the Writing Process** Chapter 31

Use your writing skills and imagination to respond to the following writing assignment. You will probably need an additional sheet of paper to complete your response.

You are a toad who has noticed that many of your friends have been run over by cars as they cross a busy highway to reach the pond where they lay their eggs.
Write a letter to your Public Works Department asking that a tunnel be built beneath the road to ensure the safe travel of your fellow toads. In your letter, list the reasons why this project is of the utmost importance.

_____

_____

_____

_____

_____

_____

_____

_____

_____

_____

_____

_____

_____

_____

_____

_____

_____

_____

_____

_____

_____

_____

## Concept Mapping                                        *Chapter 31*

The concept map below has been started for you. Add the key facts and concepts for each section of the chapter to this partial concept map. When you are done, you will have a concept map for the entire chapter.

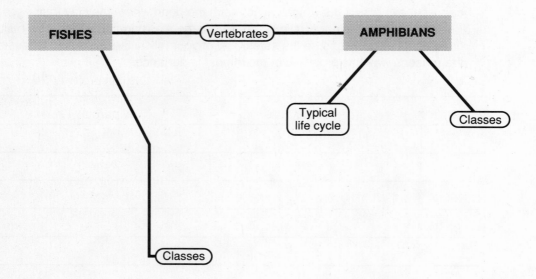

STUDY
GUIDE

| Section 32–1 | **Reptiles** | *(pages 707–719)* |

## SECTION REVIEW

Reptiles are vertebrates that have efficient lungs, a scaly waterproof skin, and amniotic eggs—adaptations that enable them to live their entire life out of water. In the first part of this section you learned about the distinguishing characteristics of reptiles. You also read a brief overview of reptile evolution.

In the second part of this section you examined the ways in which reptiles carry out their essential life functions. You learned that the lungs, brain, circulatory system, and musculoskeletal system are better developed in reptiles than in amphibians. Recall that many reptiles eliminate nitrogenous wastes as uric acid in their urine, an adaptation that helps conserve water. Reptile reproduction is also adapted to life on land: Fertilization is internal, and the egg is protected by a shell. Eggs complete their development outside the mother's body in oviparous species. In ovoviviparous species, eggs develop and hatch inside the mother's body.

In the final part of this section you studied the four living orders of reptiles: tuataras, lizards and snakes, crocodilians, and turtles. You also learned about the ways in which reptiles fit into the world.

### Identifying Structures: Building Vocabulary Skills

Examine the accompanying illustration of an amniotic egg. In the spaces provided, identify each part of the egg and write a definition for each part.

**a.** _____

_____

_____

_____

**b.** _____

_____

_____

_____

**c.** _____

_____

_____

_____

d. _____

_____

_____

_____

e. _____

_____

_____

_____

f. _____

_____

_____

### Form and Function in Reptiles: Finding the Main Ideas

The accompanying table lists some of the body systems in reptiles. In the space provided, briefly describe each of these systems.

| System | Description |
| --- | --- |
| Respiratory | |
| Circulatory | |
| Excretory | |
| Skeletal | |
| Muscular | |
| Reproductive | |

### Recognizing Fact and Fiction: Using the Main Ideas

Read the following statements about reptiles. Determine whether each statement is fact or fiction. Indicate your answer by writing the word "fact" or "fiction" on the line provided.

_____ **1.** When you touch a reptile, it feels slimy.

_____ **2.** All reptiles are carnivorous.

_____ **3.** Some species of snakes have only one lung.

_____ **4.** Turtles have better color vision than do humans.

_____ **5.** Snakes have an excellent sense of hearing.

_____ **6.** Many snakes detect the presence of a human by the body heat that the human produces.

_____ **7.** The tongue of a snake aids it in smelling.

_____ **8.** All species of snakes are poisonous.

_____ **9.** The largest lizard is the Komodo dragon.

_____ **10.** The backbone of the turtle is fused to its shell.

_____ **11.** Sea turtles provide much parental care for their young.

_____ **12.** Snakes and lizards belong to the same order of reptiles.

_____ **13.** The tuatara has a "third eye" used to detect changes in day length.

_____ **14.** Most snakes are aggressive toward humans.

_____ **15.** More Americans die from bee stings each year than from snake bites.

### Concept Mapping

The construction of and theory behind concept mapping are discussed on pages vii–ix in the front of this Study Guide. Read those pages carefully. Then consider the concepts presented in Section 32–1 and how you would organize them into a concept map. Now look at the concept map for Chapter 32 on page 315. Notice that the concept map has been started for you. Add the key facts and concepts you feel are important for Section 32–1. When you have finished the chapter, you will have a completed concept map.

## SECTION REVIEW

In this section you learned that the control of body temperature is very important for animals, particularly in habitats where temperature varies with the time of day and with the season.

Modern reptiles are called ectotherms; they obtain heat for metabolic activities from their environment. Birds and mammals are called endotherms; they generate heat inside their bodies. The first terrestrial vertebrates were probably ectotherms.

Endothermy and ectothermy are survival strategies that have advantages and disadvantages, depending on the type of environment.

### Comparing Survival Strategies: Interpreting the Main Ideas

Complete the following chart by listing two advantages and two disadvantages of ectothermy and endothermy.

| Strategy | Advantages | Disadvantages |
|---|---|---|
| Ectothermy | | |
| Endothermy | | |

### Concept Mapping

The construction of and theory behind concept mapping are discussed on pages vii–ix in the front of this Study Guide. Read those pages carefully. Then consider the concepts presented in Section 32–2 and how you would organize them into a concept map. Now look at the concept map for Chapter 32 on page 315. Notice that the concept map has been started for you. Add the key facts and concepts you feel are important for Section 32–2. When you have finished the chapter, you will have a completed concept map.

## SECTION REVIEW

In this section you learned about birds, which evolved from ancient reptiles during the Jurassic Period. Birds can be defined as oviparous endothermic reptilelike vertebrates that have feathers, two legs, and two front limbs that are modified into wings.

Because they have high metabolic rates, birds need to eat large amounts of food. The beak and feet of a bird are often highly specialized for acquiring food. The digestive system of many birds includes a crop and a gizzard. Undigestible materials, nitrogenous wastes in the form of uric acid, and reproductive materials pass through an organ called the cloaca.

Birds have a number of air sacs attached to the lungs. These air sacs help make the respiratory system more efficient than those of other vertebrates. Oxygen and other materials in the bird's blood are transported via a double-loop circulatory system powered by a four-chambered heart.

The brain and sense organs of birds are highly developed. This helps make birds capable of complex movement and complicated behaviors. Some of the most interesting bird behaviors are associated with courtship and mating. Recall that fertilization in birds is internal and that the eggs are incubated outside the mother's body. Some birds are able to take care of themselves as soon as they hatch. Others are blind, featherless, and helpless when they hatch. Such young birds require much care from their parents.

### Identifying Structures: Building Vocabulary Skills

Examine the accompanying illustration of a bird. In the spaces provided on the following page, identify each part of the bird, briefly describe its function, and name the essential life function with which it is involved.

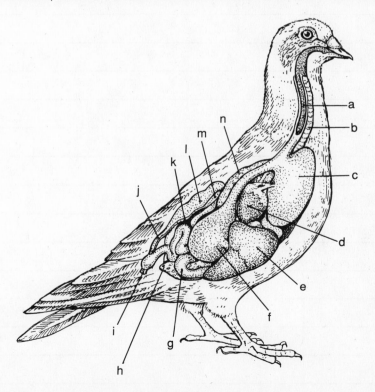

a. _____

_____

b. _____

_____

c. _____

_____

d. _____

_____

e. _____

_____

_____

f. _____

_____

g. _____

h. _____

_____

i. _____

_____

_____

j. _____

_____

k. _____

_____

l. _____

_____

m. _____

_____

n. _____

_____

### Bird Adaptations: Applying the Main Ideas

The accompanying illustration shows five species of Hawaiian honeycreepers. These birds evolved from a single ancestral species that is thought to have arrived in Hawaii several million years ago. Examine the illustration, then answer the following questions.

**1.** Why are the shapes of these birds' beaks different from one another?

_____

_____

**2.** What kind of food would you expect bird D to eat? Explain. _____

_____

**3.** What kind of food would you expect bird B to eat? Explain. _____

_____

_____

**4.** What kind of food would you expect bird C to eat? Explain. _____

_____

_____

**5.** Bird E is thought to be an important pollinator of certain flowers.

  **a.** How does bird E pollinate flowers?_____

_____

_____

**b.** What does bird E eat? _____

_____

_____

**6.** Bird A has been observed hopping around on dead trees with its beak wide open. It hammers on the tree with the thick lower part of its beak, then pokes at the tree with the thin curved upper part of its beak. Explain this behavior.

_____

_____

_____

**7.** Of the 40 known species of Hawaiian honeycreepers, 14 (including bird D) are extinct and 15 (including birds A, B, C, and E) are endangered. How might these birds' feeding

habits have contributed to their decline? _____

_____

_____

_____

**Analyzing Data**

The graph below shows the change in eggshell thickness of two types of insect-eating birds over a period of years. This period includes the years 1945–1947, when DDT, a pesticide employed for insect control, was introduced into general use.

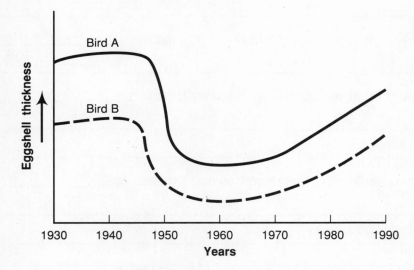

**1.** During which period of time did eggshell thickness sharply decrease?

_____

_____

**2.** What effect does DDT appear to have on eggshell thickness?

_____

_____

_____

**3.** DDT was sprayed over the land to control insects. It is believed to interfere with a bird's ability to metabolize calcium. What is the most likely explanation for how this

pesticide came to affect the eggshells of birds A and B? _____

_____

_____

_____

_____

**4.** Why might a decrease in eggshell thickness result in smaller population sizes of birds

A and B? _____

_____

_____

_____

**5.** What is the most likely explanation for the increase in eggshell thickness seen in birds

A and B in the 1970s and 1980s? _____

_____

_____

_____

_____

## Concept Mapping

The construction of and theory behind concept mapping are discussed on pages vii–ix in the front of this Study Guide. Read those pages carefully. Then consider the concepts presented in Section 32–3 and how you would organize them into a concept map. Now look at the concept map for Chapter 32 on page 315. Notice that the concept map has been started for you. Add the key facts and concepts you feel are important for Section 32–3. When you have finished the chapter, you will have a completed concept map.

## Using the Writing Process                              *Chapter 32*

Use your writing skills and imagination to respond to the following writing
assignment. You will probably need an additional sheet of paper to
complete your response.

Imagine that you have been transformed into the reptile of your choice. Write a
short story about your experiences as a reptile.

_____

_____

_____

_____

_____

_____

_____

_____

_____

_____

_____

_____

_____

_____

_____

_____

_____

_____

_____

_____

_____

_____

_____

## Concept Mapping                                              *Chapter 32*

The concept map below has been started for you. Add the key facts and concepts for each section of the chapter to this partial concept map. When you are done, you will have a concept map for the entire chapter.

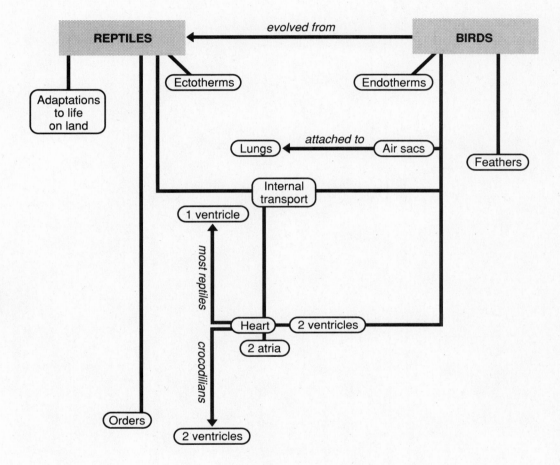

S T U D Y
G U I D E

CHAPTER **33**
*Mammals*

| Section 33–1 | **Mammals** | *(pages 737–745)* |

### SECTION REVIEW

In this section you were introduced to the group of endothermic vertebrates known as mammals. You learned about the general characteristics of mammals and about a few interesting adaptations of particular mammals.

The single most important characteristic of mammals is that females have mammary glands. The mammary glands produce milk that nourishes the young after they are born.

Mammals also have a variety of other distinguishing characteristics. Their bodies are insulated by various combinations of fur, hair, and subcutaneous fat. And many mammals have sweat glands that help cool the body. Almost all mammals have several kinds of teeth, which are specialized for processing

food in different ways. All mammals have lungs and a breathing muscle called the diaphragm. They have a double-loop circulatory system and a four-chambered heart. Mammals have the most highly developed kidneys and brains of all vertebrates.

Mammals are divided into three groups according to their method of reproduction. Monotreme embryos complete their development in a leathery-shelled egg that develops outside the mother's body. Marsupial embryos are born at a very early stage of development and complete their development inside the mother's marsupium, or pouch. Placental embryos complete their development inside the mother's uterus.

### The Mammalian Brain: Building Vocabulary Skills

The accompanying diagram shows the structure of a typical mammal's brain. In the space provided, identify each part of the brain and describe its function.

**a.** _____

_____

_____

_____

**b.** _____

_____

_____

**c.** _____

_____

_____

_____

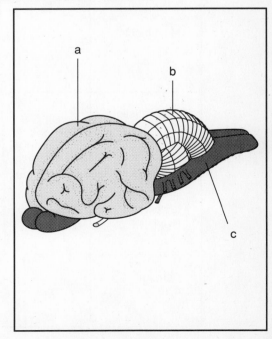

■ **Summarizing Information: Finding the Main Ideas**

The following table lists a number of functions in animals. Complete the table by briefly noting what you have learned about each of these functions in mammals.

| Function | Characteristics of Mammals |
|---|---|
| Feeding | |
| Respiration | |
| Internal transport | |
| Excretion | |
| Response | |
| Movement | |
| Reproduction | |
| Body temperature regulation | |
| Parental care | |

### Studying Dolphin Migration: Analyzing Data

A certain species of dolphin migrates every year from northern waters to waters just off the coast of the United States. In these southern waters, the dolphins give birth to young and breed. While investigating the migratory behavior of these mammals, scientists counted the dolphins in the breeding area every month. The accompanying graph shows the results of the monthly count.

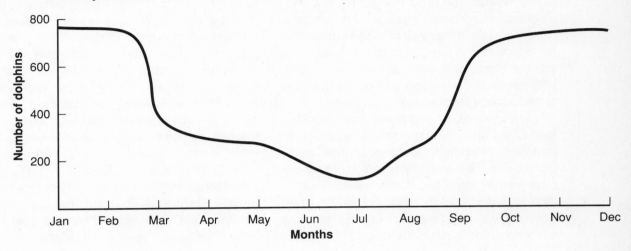

1. During which months are the most dolphins in the breeding area?

_____

2. During which months are the fewest dolphins in the breeding area?

_____

3. During which months is the breeding season for the dolphins likely?

_____

4. What is the probable reason that the dolphins migrate south to give birth and to breed?

_____

_____

5. What does a newborn dolphin eat? How does it get its food?

_____

### Concept Mapping

The construction of and theory behind concept mapping are discussed on pages vii–ix in the front of this Study Guide. Read those pages carefully. Then consider the concepts presented in Section 33–1 and how you would organize them into a concept map. Now look at the concept map for Chapter 33 on page 324. Notice that the concept map has been started for you. Add the key facts and concepts you feel are important for Section 33–1. When you have finished the chapter, you will have a completed concept map.

## Section 33–2  Important Orders of Living Mammals  *(pages 746–751)*

### SECTION REVIEW

Living mammals are divided into approximately twenty orders. (The exact number is still a matter of scientific debate.) Important characteristics used to classify a mammal include its method of reproduction; the number of bones in its skull; and the structure, number, and kinds of teeth it possesses. In this section you were introduced to the fourteen major orders of mammals.

Monotremes, or egg-laying mammals, belong to the order Monotremata. Marsupials, or pouched mammals, belong to the order Marsupialia. Placental mammals make up the other mammalian orders. Order Insectivora contains small mammals that typically have long, narrow mobile snouts and feed on invertebrates such as insects. Bats belong to the order Chiroptera. Members of the order Edentata lack incisors, canines, and premolars, and may be entirely without teeth. Rodents, or members of the order Rodentia, are gnawing mammals. The order Lagomorpha, whose name means rabbit-shaped, contains rabbits, hares, and pikas. The order Carnivora consists of meat-eaters, such as cats, dogs, bears, raccoons, weasels, hyenas, and seals. Whales, porpoises, and dolphins belong to the order Cetacea. Manatees and sea cows belong to the order Sirenia. The order Artiodactyla contains even-toed hoofed mammals. Elephants belong to the order Proboscidea. The order Primates contains lemurs, monkeys, apes, and humans.

### Classifying Mammals: Using the Main Ideas

The accompanying illustrations show a number of different kinds of mammals. In the space provided, write the name of the order to which the mammals in each illustration belong.

Long-beaked echidna

Armadillo

Elephant

Jack rabbit

Moose

Hedgehog

_____ | _____

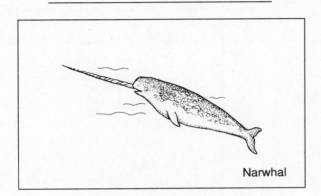

Narwhal

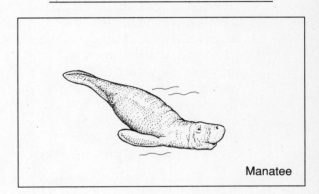

Manatee

_____ | _____

Wolf

Horse

_____ | _____

Opossum

Bat

_____ | _____

## Concept Mapping

The construction of and theory behind concept mapping are discussed on pages vii–ix in the front of this Study Guide. Read those pages carefully. Then consider the concepts presented in Section 33–2 and how you would organize them into a concept map. Now look at the concept map for Chapter 33 on page 324. Notice that the concept map has been started for you. Add the key facts and concepts you feel are important for Section 33–2. When you have finished the chapter, you will have a completed concept map.

Use your writing skills and imagination to respond to the following writing assignment. You will probably need an additional sheet of paper to complete your response.

Write a humorous dialogue in which a reporter interviews a mammal of your choice. In the course of the interview, try to convey as much information as you can about form and function in the mammal.

_____

_____

_____

_____

_____

_____

_____

_____

_____

_____

_____

_____

_____

_____

_____

_____

_____

_____

_____

_____

_____

_____

## Concept Mapping *Chapter 33*

The concept map below has been started for you. Add the key facts and concepts for each section of the chapter to this partial concept map. When you are done, you will have a concept map for the entire chapter.

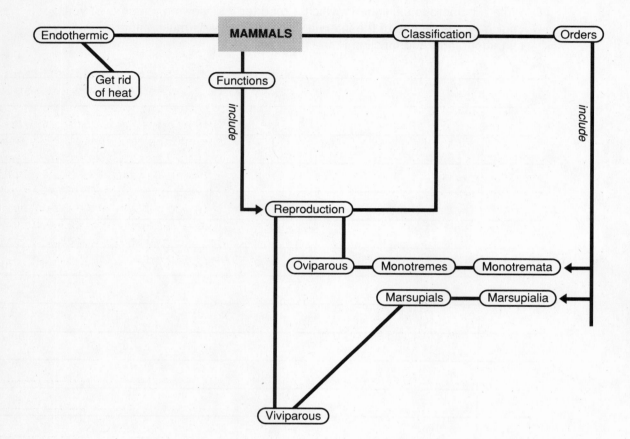

© Prentice-Hall, Inc.

S T U D Y
G U I D E

CHAPTER **34**
*Humans*

| Section 34–1 | **Primates and Human Origins** | *(pages 757–759)* |

### SECTION REVIEW

In this section you learned that humans (*Homo sapiens*) evolved from common ancestors shared with other living primates such as chimpanzees and apes.

Primates share certain characteristics: flat faces, reduced snouts, eyes that face forward and allow for binocular vision, flexible fingers and toes, arms that can rotate in broad circles around the shoulders, and a large cerebrum.

Very early in primate history, prosimians branched off from the main primate line. Living prosimians include lemurs and lorises. Anthropoid primates include monkeys, apes, and humans.

Two anthropoid branches separated around 45 million years ago when the continents shifted and were no longer connected by land bridges. One anthropoid branch evolved into the New World monkeys, which are primarily tree dwellers and, in general, have prehensile tails. The other anthropoid branch evolved into the Old World monkeys, which include baboons and the great apes.

The great apes, also called hominoids, include chimpanzees, orangutans, gorillas, and *Homo sapiens*.

### Identifying Characteristics: Finding the Main Ideas

In the spaces provided below, list seven characteristics of the order Primates.

1. _____

    _____

2. _____

    _____

3. _____

    _____

4. _____

    _____

5. _____

    _____

6. _____

    _____

7. _____

    _____

### Comparing Primates

The chart below lists the different evolutionary lines of primates. In the spaces provided, describe each group of primates and list an example of each group.

| Primate Line | | | Characteristics | Example |
|---|---|---|---|---|
| Prosimians | | | | |
| Anthropoids | New World monkeys | | | |
| | Old World monkeys | Monkeys | | |
| | | Hominoids | | |

### Concept Mapping

The construction of and theory behind concept mapping are discussed on pages vii–ix in the front of this Study Guide. Read those pages carefully. Then consider the concepts presented in Section 34–1 and how you would organize them into a concept map. Now look at the concept map for Chapter 34 on page 332. Notice that the concept map has been started for you. Add the key facts and concepts you feel are important for Section 34–1. When you have finished the chapter, you will have a completed concept map.

## SECTION REVIEW

In this section you learned that the hominoid (great apes) line gave rise to a small group of species called hominids. Hominids were omnivores that ate both meat and vegetable foods.

Hominid adaptations include changes in the spinal column, hip bones, and leg bones that allowed bipedal (two-foot) locomotion. Hominids also have a much larger brain than other primates.

The first recognized hominids were the australopithecines, all of which walked erect. *Australopithecus* species include *A. boisei, A. robustus, A. afarensis,* and *A. africanus.*

The first species to be classified in the genus *Homo* was *Homo habilis,* which means handy man. *Homo habilis* used stone tools and may have hunted for meat or scavenged from the kills of other carnivores.

*Homo habilis* was replaced by *Homo erectus,* which spread throughout much of the world around 1 million years ago. *Homo erectus* was an accomplished toolmaker and could also use fire.

Although the first *Homo sapiens* evolved around 500,000 years ago, the first group to resemble modern humans seems to have evolved around 150,000 years ago and is called *Homo sapiens neanderthalensis.* Over time, Neanderthals were replaced by Cro-Magnon, who are classified as modern humans, or *Homo sapiens sapiens.*

## Identifying Relationships

A positive-feedback system is a series of events in which two or more changes in an organism complement one another. As one component of the system develops, it reinforces another component. In this activity you will examine how the development of the size of the brain and its capacity as well as the development of small canine teeth might have worked together to advance the development of humans. Figure 1 shows a positive-feedback system. Use the descriptions that follow to label the positive-feedback mechanisms shown in each step of Figure 1. The first two have been done for you.

**a.** Hominids had small canine teeth and had to be able to make tools and weapons.

**b.** Increased brain size made it possible to pursue more efficient hunting.

**c.** Bipedalism freed the hands and arms for throwing weapons.

**d.** Toolmaking required increased brain size and complexity.

**e.** Hunting required bipedalism.

**f.** Carrying weapons, tools, food, and offspring required bipedalism.

**g.** Bipedalism made it possible to carry weapons, tools, food, and offspring.

**h.** Increased brain size made toolmaking possible.

**i.** Hominids had small canine teeth and had to be able to throw weapons to protect themselves.

**j.** Toolmaking made hunting possible.

**k.** The most effective hunters were those who had more complex brains.

**l.** Throwing weapons required hands free of other objects.

**m.** Hunting required toolmaking.

**n.** The capability of throwing weapons made large canine teeth unnecessary for protection.

**o.** Bipedalism made hunting possible.

**p.** Toolmaking required bipedalism; hands and arms needed to be free.

**Figure 1**

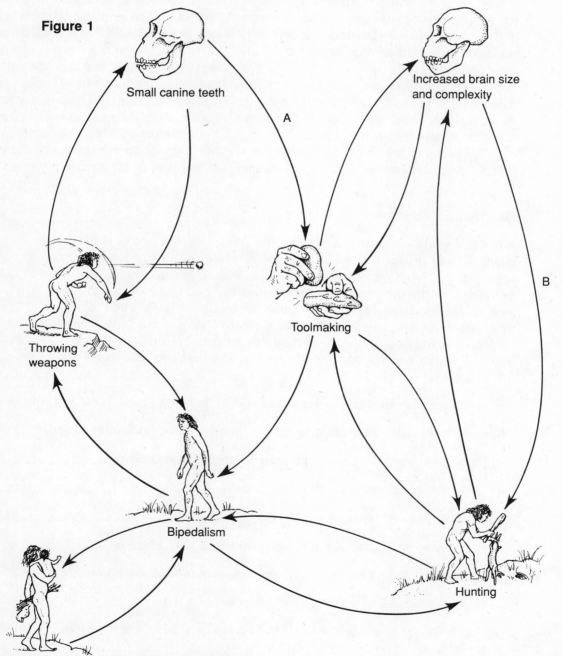

Small canine teeth

Increased brain size and complexity

A

Throwing weapons

Toolmaking

B

Bipedalism

Hunting

Carrying weapons, tools, food, offspring

### Interpreting Diagrams

Look at the diagrams and answer the questions that follow.

H. sapiens

H. erectus

A. robustus

H. habilis

A. africanus

A. afarensis

**Johanson's scheme**

H. sapiens — Present

H. erectus

A. robustus

H. habilis

A. africanus

A. afarensis

?

— 4.5

**Leakey's scheme**

Millions of years ago

1. What do the two schemes illustrate?

_____

_____

_____

2. In each scheme, identify the oldest species and the most recent species.

_____  _____

_____

_____

_____

3. What is the evolutionary relationship among the three species in the genus *Homo?*

_____

_____

_____

4. What point in each scheme most closely represents the present time?

_____

_____

_____

5. What does the question mark in Leakey's

scheme represent?_____

_____

_____

_____

6. What would allow Leakey to substitute an actual name for the question mark?

_____

_____

_____

## Comparing Anthropoid Characteristics: Finding the Main Ideas

Place an X in the appropriate box if the characteristic is demonstrated by the species.

|  | Chimpanzees and Gorillas | Australopithe-cus africanus | Homo Habilis | Homo erectus | Neanderthal | Modern Humans |
|---|---|---|---|---|---|---|
| Two-legged walking |  |  |  |  |  |  |
| Large brain |  |  |  |  |  |  |
| Made and used tools |  |  |  |  |  |  |
| Used fire for warmth and cooking |  |  |  |  |  |  |
| Buried their dead |  |  |  |  |  |  |
| Communicated by use of a spoken language |  |  |  |  |  |  |

## Concept Mapping

The construction of and theory behind concept mapping are discussed on pages vii–ix in the front of this Study Guide. Read those pages carefully. Then consider the concepts presented in Section 34–2 and how you would organize them into a concept map. Now look at the concept map for Chapter 34 on page 332. Notice that the concept map has been started for you. Add the key facts and concepts you feel are important for Section 34–2. When you have finished the chapter, you will have a completed concept map.

## Using the Writing Process                                    *Chapter 34*

Use your writing skills and your imagination to respond to the following writing assignment. You will probably need an additional sheet of paper to complete your response.

Imagine you could go back in time to present one tool from modern Earth to *Homo habilis,* the handy man. What tool would you select? Explain your answer in words or drawings—or even a video.

_____

_____

_____

_____

_____

_____

_____

_____

_____

_____

_____

_____

_____

_____

_____

_____

_____

_____

_____

## Concept Mapping

The concept map below has been started for you. Add the key facts and concepts for each section of the chapter to this partial concept map. When you are done, you will have a concept map for the entire chapter.

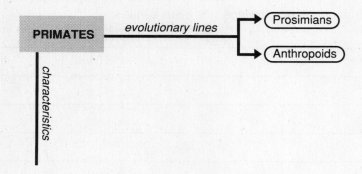

STUDY
GUIDE

| Section 35-1 | **Elements of Behavior** | *(pages 771–775)* |

### SECTION REVIEW

In this section you learned that the behavior of an animal is just as important to its survival and reproduction as any of its physical characteristics. Instincts are behaviors that are built in to an animal's nervous system and cannot be changed during the animal's lifetime.

Learning is the way in which an animal changes its behavior as a result of experience. An animal learns in several different ways. Habituation is a decrease in response to an unimportant stimulus. Classical conditioning occurs when an animal makes a mental connection between a stimulus and some kind of good or bad event.

In operant conditioning, or trial-and-error learning, an animal learns to behave in a certain way to receive a reward or to avoid punishment. In insight learning, an animal applies something it has learned to a new situation without a period of trial and error.

Some behaviors, although primarily instinctive, cannot occur without some learning on the part of the animal. In the process of imprinting, for example, newborn ducks and geese combine their natural instinct to follow their mother with an image obtained by experience.

### Relating Concepts: Finding the Main Ideas

The chart below lists some of the different types of animal behavior. In the space provided, give a description and one example of each type of behavior.

| Behavior | Description | Example |
|---|---|---|
| Instinct | | |
| Habituation | | |
| Classical conditioning | | |
| Operant conditioning | | |
| Insight learning | | |

### Reading Maps

Each of the following maps shows the yearly migration route of a marine mammal. Map A shows the route of a fur seal; map B shows the route of the gray whale. Answer the questions based on the maps.

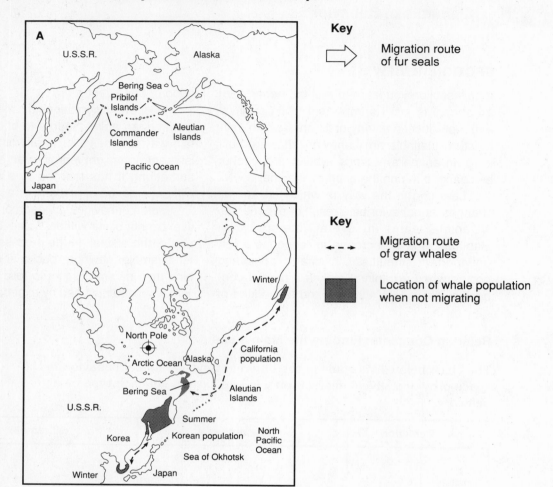

**Key**

⇨ Migration route of fur seals

**Key**

◄-- --► Migration route of gray whales

■ Location of whale population when not migrating

**1.** How do the maps differ in terms of geographic arrangement? _____

_____

_____

**2.** What advantages are there to the orientation of each map? _____

_____

_____

_____

**3.** What do the arrows represent on each map? _____

_____

_____

4. How many populations are represented in map A? In map B? What are the similarities between the migration routes of the populations in map A? What are the differences?

_____

_____

_____

_____

_____

5. What prevents competition between the populations represented in map B?

_____

_____

_____

_____

6. Which map could show the migration routes of both the fur seal and the gray whale? Explain

your answer. _____

_____

_____

_____

7. How would you indicate the migration routes and the summer and winter ranges of both

seal and whale populations on the same map? _____

_____

_____

_____

### Concept Mapping

The construction of and theory behind concept mapping are discussed on pages vii–ix in the front of this Study Guide. Read those pages carefully. Then consider the concepts presented in Section 35–1 and how you would organize them into a concept map. Now look at the concept map for Chapter 35 on page 340. Notice that the concept map has been started for you. Add the key facts and concepts you feel are important for Section 35–1. When you have finished the chapter, you will have a completed concept map.

## Communication: Signals for Survival

*(pages 775–777)*

### SECTION REVIEW

In this section you learned that communication is the passing of information from one animal to another. Animals use many varied techniques to communicate with each other.

Animals communicate with each other for many reasons. For example, animals communicate in order to choose a mate, to transmit information about the location and availability of food, and to warn others about potential dangers.

The way in which animals communicate is limited only by the kinds of stimuli their senses can detect. Methods of communication include visual signals, sound signals, chemical signals, and electrical signals.

### Comparing Honeybee and Human Behavior: Applying Concepts

Read the paragraphs below. Then complete the chart on the following page by comparing the honeybee community and the human community. Focus special attention on the role that communication plays in each group.

Human communication is usually considered to be the most sophisticated and efficient in the animal kingdom. Yet, a honeybee colony is as well organized and as productive as any group of humans. While reading the next paragraph, use your knowledge of human behavior and insect behavior to analyze honeybee behavior and to compare the behavior of honeybees to that of humans.

For social animals, the ability to communicate is essential for the survival of the group. Honeybees communicate mostly through taste and smell. Bees transmit information to one another through the transfer of food. Young larvae are fed special food that is produced by glands in nurse bees. Queen bees and older larvae also receive this secretion. Food-gathering bees bring nectar that they have collected in their crops and feed it to house bees. House bees transfer the nectar to other house bees by using their crops and mouths. All of these food exchanges among honeybees make up a chemical communication system. The nectar sample that a food-gathering bee brings back to the colony provides information about the kind of food that it found. This method of communication ensures that other members of the colony will search for the same kind of food and that all members of the colony will be eating the same kind of diet. The diet produces similar body odors among the members of the colony. The transfer of food from one bee to another also passes along pheromones, the chemical messengers produced by the bees that provide additional information that the bees can use to identify members of a colony. The food sources of a colony are distinct enough from the food sources of other colonies in the area to prevent any confusion on the part of the members of the two colonies. Thus, the odor of one colony is never identical to that of any other colony. Members of a colony are able to identify themselves and to recognize the differences between their members and members of neighboring colonies.

| Task | Honeybees | Humans |
|------|-----------|--------|
| Obtaining food | | Use higher brain functions to grow own food; communicate through speech and body language in hunting and agriculture |
| Feeding of young | | |
| Feeding of adult | Food-gathering bees regurgitate nectar to house bees; house bees transfer food to one another by using crops and mouths. | |
| Obtaining shelter | | Adults select homes as individuals or as members of a group, taking factors such as work and school into consideration. |
| Division of labor | | Varies in each family and in each community according to needs and customs |
| Communication | | |
| Hormones | | Elaborate hormone system used by individuals to maintain body functions; not used to communicate with individuals. |

## Concept Mapping

The construction of and theory behind concept mapping are discussed on pages vii–ix in the front of this Study Guide. Read those pages carefully. Then consider the concepts presented in Section 35–2 and how you would organize them into a concept map. Now look at the concept map for Chapter 35 on page 340. Notice that the concept map has been started for you. Add the key facts and concepts you feel are important for Section 35–2. When you have finished the chapter, you will have a completed concept map.

**Evolution of Behavior** *(pages 778-779)*

### SECTION REVIEW

In this section you learned that the DNA in an animal's genes codes for certain behaviors as well as for its physical characteristics. Like physical characteristics, variations in the genetic material that codes for behavior can be inherited if the behavior contributes to the animal's survival.

In some species, the fitness of an individual is not affected by cooperating with other members of its species. However, when social behavior offers great survival advantages, natural selection favors the evolution of such behaviors.

#### Applying Concepts: Finding the Main Ideas

In the space provided below, describe four specific examples of animal behaviors that are controlled by an animal's genetic coding.

1. _____
   _____
   _____

2. _____
   _____
   _____

3. _____
   _____
   _____

4. _____
   _____
   _____

#### Concept Mapping

The construction of and theory behind concept mapping are discussed on pages vii–ix in the front of this Study Guide. Read those pages carefully. Then consider the concepts presented in Section 35–3 and how you would organize them into a concept map. Now look at the concept map for Chapter 35 on page 340. Notice that the concept map has been started for you. Add the key facts and concepts you feel are important for Section 35–3. When you have finished the chapter, you will have a completed concept map.

Name _____ Class _____ Date _____

Use your writing skills and imagination to respond to the following
writing assignment. You will probably need an additional sheet of
paper to complete your response.

Changing family structures, television and other forms of entertainment,
as well as the pace of life have changed the amount of time many people
spend in meaningful communication. Write a magazine article on the
possible effect this trend might have on society if it continues.

_____

_____

_____

_____

_____

_____

_____

_____

_____

_____

_____

_____

_____

_____

_____

_____

_____

_____

_____

_____

_____

_____

## Concept Mapping *Chapter 35*

The concept map below has been started for you. Add the key facts and concepts for each section of the chapter to this partial concept map. When you are done, you will have a concept map for the entire chapter.

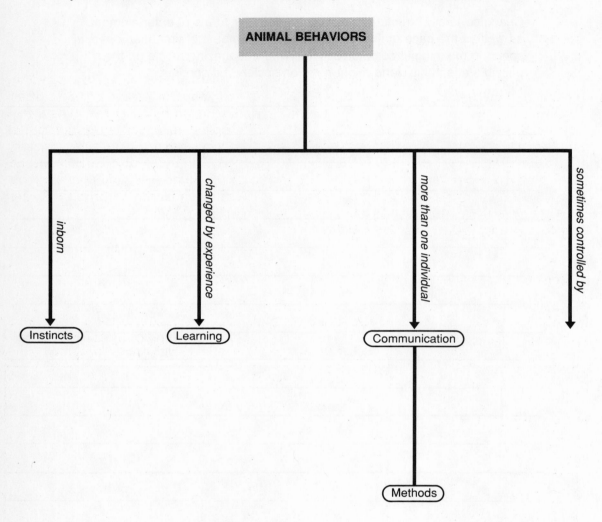

**ANIMAL BEHAVIORS**

*inborn*

*changed by experience*

*more than one individual*

*sometimes controlled by*

Instincts

Learning

Communication

Methods

S T U D Y
G U I D E

| Section 36–1 | **Evolution of the Vertebrates** | *(pages 785–789)* |

## SECTION REVIEW

In this section you learned that our best understanding of the evolutionary relationships between vertebrate groups can be shown as a phylogenetic tree.

In divergent evolution, related evolutionary lines become more dissimilar as they are subjected to different forces of natural selection. Divergent evolution is also known as adaptive radiation. In convergent evolution, evolutionary lines that are subjected to similar forces of natural selection become more similar to one another as they evolve.

Ectotherms must obtain the heat they need from their environment. They typically rely primarily on behavior to regulate their body temperature. Endotherms generate all the heat they need through metabolic activity. They typically rely on physiological mechanisms to regulate their body temperature. They also use a number of behaviors to prevent overheating.

### Comparing Evolutionary Complexity: Finding the Main Ideas

Complete the following chart.

| Animal | Vertebrate Group | Three Other Examples of Animals in This Group |
|---|---|---|
| a. | | |
| b. | | |
| c. | | |
| d. | | |

### Comparing Evolutionary Trends: Building Vocabulary Skills

The chart below lists two general evolutionary trends. In the spaces provided, describe and give examples of each trend.

| Evolutionary Trend | Description | Examples |
|---|---|---|
| Convergent | | |
| Divergent | | |

### Concept Mapping

The construction of and theory behind concept mapping are discussed on pages vii–ix in the front of this Study Guide. Read those pages carefully. Then consider the concepts presented in Section 36–1 and how you would organize them into a concept map. Now look at the concept map for Chapter 36 on page 348. Notice that the concept map has been started for you. Add the key facts and concepts you feel are important for Section 36–1. When you have finished the chapter, you will have a completed concept map.

## Section 36–2  Form and Function in Vertebrates                    *(pages 789–799)*

### SECTION REVIEW

In this section you learned that as you move through the vertebrate classes from fishes to mammals, organ systems tend to become increasingly complex. In more primitive vertebrates, the limbs stick out from the sides of the body. In more advanced vertebrates, the limbs tend to be positioned directly beneath the body.

The digestive systems of vertebrates are adapted to a number of different foods and methods of feeding.

Some vertebrates use gills for respiration; others use lungs. Lungs increase in efficiency as you move from amphibians to reptiles to mammals. Birds have the most advanced respiratory system of all vertebrates.

Vertebrates that have a single-loop circulatory system also have a two-chambered heart. Double-loop circulatory systems are associated with lungs. As vertebrates with lungs evolved,

the separation of the two loops of the circulatory system improved. Frogs and toads have a three-chambered heart. Most reptiles have a three-chambered heart that has partial partition in the ventricle. Birds, mammals, and crocodilians have a four-chambered heart.

Most fishes and aquatic amphibians excrete nitrogenous wastes in the form of ammonia. Mammals and most cartilaginous fishes excrete urea. Birds and reptiles excrete uric acid.

As you move through the vertebrate classes from fishes to mammals, the relative size and complexity of the cerebrum and cerebellum increase.

Primitive vertebrates tend to have external fertilization; more advanced vertebrates tend to have internal fertilization. Vertebrates may be oviparous, ovoviviparous, or viviparous.

### Identifying Internal Structures: Building Vocabulary Skills

The drawings below represent the digestive systems of different classes of vertebrates. Identify each of the numbered parts in the spaces provided on the following page.

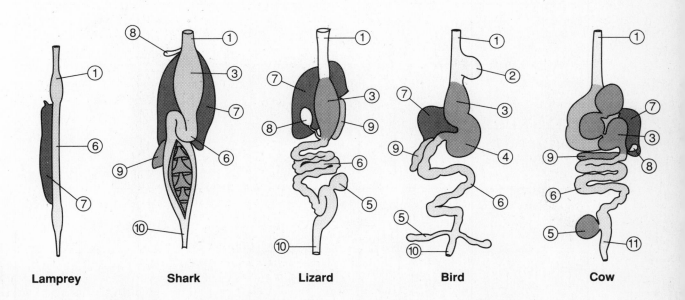

Lamprey          Shark          Lizard          Bird          Cow

## Comparing Vertebrate Characteristics

Using the following list of characteristics, fill in the chart below.

Hops; has four legs
Lays soft-shelled eggs
 on land
Insects, worms, berries,
 seeds
Deserts; forests
Crawls on belly
Queen lays thousands of
 eggs
Lizards, insects, small
 snakes, rodents
Swims using fins
Ponds, marshes, streams
Trees
Lays eggs in water

Has six legs; walks; flies
Walks and runs on four legs
Small mollusks, worms,
 small crabs
Produces live young
Wooded or bushy areas
Pollen and nectar
Lays eggs in water
Flies; walks on two legs
Fields; flower gardens
Lays hard-shelled eggs in a
 nest
Rats, mice, rabbits, squirrels
Salt water or fresh water
Insects; spiders

| Animal | Movement | Reproduction | Food | Habitat |
|--------|----------|--------------|------|---------|
| Bullfrog | | | | |
| Blackbird | | | | |
| Flounder | | | | |
| Honeybee | | | | |
| Coral snake | | | | |
| Red fox | | | | |

### Analyzing Vertebrate Adaptations

Humans and other vertebrate land animals are adapted for living in air. Fishes are adapted for living in water. Fill in the chart below by comparing the adaptations that enable land vertebrates and fishes to perform various life processes.

| Life Process | Vertebrate Land Animals | Fishes |
|---|---|---|
| Getting oxygen | | |
| Moving | | |
| Excretion | | |
| Internal transport | | |

1. Explain why fishes need to be more streamlined than land animals.

_____

_____

_____

_____

2. Fishes are covered with a slimy material that is produced by glands in the skin. Suggest

a reason for this covering. _____

_____

_____

_____

**3.** When a puffer fish is approached by a predator, it puffs itself up, causing its spines to stick out from its body. What land animal uses a similar form of protection?

_____

_____

**4.** The moray eel is a fish that swims by wriggling its body. What land animal moves in a

similar way?_____

_____

### Identifying Internal Structures: Building Vocabulary Skills

The drawing below represents the reproductive systems of several classes of vertebrates. Identify each of the numbered parts.

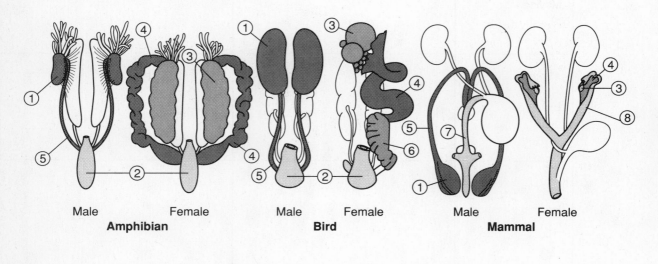

Male      Female       Male    Female       Male      Female
**Amphibian**         **Bird**          **Mammal**

| | |
|---|---|
| 1. _____ | 5. _____ |
| 2. _____ | 6. _____ |
| 3. _____ | 7. _____ |
| 4. _____ | 8. _____ |

### Concept Mapping

The construction of and theory behind concept mapping are discussed on pages vii–ix in the front of this Study Guide. Read those pages carefully. Then consider the concepts presented in Section 36–2 and how you would organize them into a concept map. Now look at the concept map for Chapter 36 on page 348. Notice that the concept map has been started for you. Add the key facts and concepts you feel are important for Section 36–2. When you have finished the chapter, you will have a completed concept map.

## Using the Writing Process

Use your writing skills and imagination to respond to the following writing assignment. You will probably need an additional sheet of paper to complete your response.

Here is a story that is neither ancient nor particularly sage. But it is an interesting story and something of a mystery, for the ending is still unknown.

One day a young girl found a bottle with a genie inside. In exchange for his freedom, the genie granted the girl three wishes. However, he specified that she could wish only for the improved senses of other vertebrates (to the genie, one animal with a backbone was just like any other). The young girl thought for a moment—and only a moment, as is often the habit with junior high students—and then selected the hearing ability of her dog and the smelling ability of her cat. She also wanted to be able to detect infrared radiation like a rattlesnake. Having made her choices, the young girl hurried off to school . . .

Finish the story using the girl as your main character. If you can, try to show both the advantages and the perhaps surprising disadvantages of each choice.

_____

_____

_____

_____

_____

_____

_____

_____

_____

_____

_____

_____

_____

_____

_____

_____

_____

**Concept Mapping**                                    *Chapter 36*

The concept map below has been started for you. Add the key facts and concepts for each section of the chapter to this partial concept map. When you are done, you will have a concept map for the entire chapter.

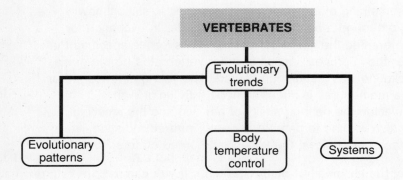

STUDY
GUIDE

| Section 37–1 | The Nervous System | (pages 809–815) |

## SECTION REVIEW

In this section you were introduced to the system that allows living things to communicate with their environment and with all the cells of their body: the nervous system. Cells called neurons carry information throughout the nervous system. This information is in the form of electrical signals called nerve impulses. Neurons come in many shapes and sizes, but they all have the same basic structure. Dendrites carry impulses from the environment or other cells in the body to the cell body of the neuron. The impulse travels through the cell body to the axon and on to the next neuron. Some of the neuron's axons are covered with a sheath of myelin. Myelin allows nerve impulses to travel more quickly along the axon because the impulse can jump from node to node on the axon instead of moving along the entire cell membrane.

Later in this section you learned that neurons can be divided into three types based on the direction in which nerve impulses move. Sensory neurons carry information gathered by the sense organs to the spinal cord and brain. Motor neurons carry information from the brain and spinal cord to the muscles and glands. Interneurons connect and allow communication between sensory and motor neurons.

In the last part of this section you read about the nerve impulse and synapse. You also learned that a nerve impulse is the flow of electrical charges along the cell membrane of a neuron. The difference in charge on either side of the cell membrane is called the resting potential of a neuron's cell membrane. As positive ions of potassium and sodium are pumped across the cell membrane, the electrical charge on either side of the cell membrane changes and the nerve impulse or action potential travels along the neuron. When the nerve impulse reaches the axon terminals at the end of the axon, the impulse must cross the synapse to the dendrites of the next neuron. Chemicals called neurotransmitters allow the impulse to travel from one neuron to the next.

### Applying Definitions: Building Vocabulary Skills

1. In the space below, draw and label a neuron. The neuron should have the following labels: axon, axon terminals, cell body, dendrite, myelin sheath, node, and nucleus.

**2.** Describe the path a nerve impulse follows as it travels along a neuron.

_____

_____

_____

_____

_____

_____

_____

_____

### Relating Concepts: Finding the Main Ideas

*Situation:* You are standing on your driveway waiting for a ride to school. Suddenly it begins to rain. As your skin gets pelted by huge cold raindrops, you run for cover and decide to wait for your ride protected by the front porch.

**1.** What is the stimulus in this situation? _____

_____

_____

_____

_____

**2.** What type of neurons carried information about the stimulus to the brain?

_____

_____

_____

_____

_____

**3.** What type of neurons carried instructions to your muscles to run for cover?

_____

_____

_____

_____

**4.** Did the stimulus in this example reach the threshold? Explain your answer.

_____

_____

_____

_____

_____

_____

**Sequencing Events: The Nerve Impulse**

Place the events described below in the correct sequence by writing the numbers 1 through 6 in the spaces provided.

_____ Impulse reaches the axon terminals.

_____ Sodium ions flow inside the cell membrane and the cell membrane is depolarized.

_____ The cell membrane of a neuron is polarized.

_____ Neurotransmitters of one neuron depolarize the cell membrane of another neuron to allow the impulse to cross the synapse.

_____ The nerve impulse flows in one direction down the axon. As the nerve impulse passes along the neuron, potassium ions move outside the cell membrane and repolarize it.

_____ Stimulus reaches the threshold and produces an impulse that activates the dendrites of the neuron.

**Concept Mapping**

The construction of and theory behind concept mapping are discussed on pages vii–ix in the front of this Study Guide. Read those pages carefully. Then consider the concepts presented in Section 37–1 and how you would organize them into a concept map. Now look at the concept map for Chapter 37 on page 362. Notice that the concept map has been started for you. Add the key facts and concepts you feel are important for Section 37–1. When you have finished the chapter, you will have a completed concept map.

## Section 37–2 | Divisions of the Nervous System | *(page 816)*

### SECTION REVIEW

In this section you learned that the nervous system is divided into two major parts: the central nervous system and the peripheral nervous system. The central nervous system, made up of the brain and spinal cord, is the control center for the body. The central nervous system is responsible for analyzing data, processing information, and relaying messages. The peripheral nervous system is made up of all the parts of the nervous system outside of the brain and spinal cord. Because the peripheral nervous system comes into contact with the environment, it provides the central nervous system with information to analyze and process. The peripheral nervous system also carries out the responses ordered by the central nervous system.

### Applying Definitions: Building Vocabulary Skills

Identify the central nervous system in the diagram below by coloring in the parts of the central nervous system with a pen or colored pencil. Label the two major parts of the central nervous system.

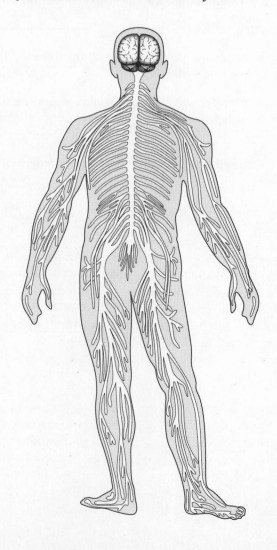

### Relating Concepts: Finding the Main Ideas

Put a C next to each item that describes the central nervous system or its functions. Put a P next to each item that describes the peripheral. If an item describes both parts of the nervous system, place both a P and a C in the blank.

1. Analyzing data needed to solve a problem _____

2. Composed of the brain and spinal cord _____

3. Gathering information about the environment _____

4. Relaying messages _____

5. Cranial and spinal nerves _____

6. Protected by bone _____

7. Acts like the processing unit of a computer _____

8. Comes into contact with the environment _____

9. Made up of neurons _____

10. Provides data to be analyzed, processed, and stored _____

### Analyzing Information: Drawing Conclusions

1. Why is it important that the central nervous system be protected by bone?

_____

_____

_____

2. How would your life be different if your peripheral nervous system did not function

properly? _____

_____

_____

_____

### Concept Mapping

The construction of and theory behind concept mapping are discussed on pages vii–ix in the front of this Study Guide. Read those pages carefully. Then consider the concepts presented in Section 37–2 and how you would organize them into a concept map. Now look at the concept map for Chapter 37 on page 362. Notice that the concept map has been started for you. Add the key facts and concepts you feel are important for Section 37–2. When you have finished the chapter, you will have a completed concept map.

| Section 37–3 | The Central Nervous System | *(pages 817–823)* |

## SECTION REVIEW

In this section you were introduced to the structure and function of the central nervous system. The central nervous system consists of the brain and spinal cord.

In the first part of this section you learned about the structure and function of the brain. The brain is the organ into which nerve impulses flow for processing and from which nerve impulses flow to control the body. The major parts of the brain are the cerebrum, cerebellum, brainstem, hypothalamus, and thalamus.

The cerebrum controls all voluntary activities and is responsible for learning, judgment, and intelligence. The cerebrum is divided into right and left hemispheres and into an outer cerebral cortex and an inner cerebral medulla.

The cerebellum coordinates the action of muscles so that the body can move effectively.

The brainstem coordinates and integrates all the information that enters the brain. The brainstem consists of the medulla oblongata, pons, and midbrain.

The thalamus and hypothalamus are located near the center of the brain, just above the brainstem. The thalamus receives sensory information and forwards it to the proper part of the cerebrum for processing. The hypothalamus is the control center for hunger, thirst, fatigue, anger, and body temperature.

In the second part of this section you learned that some regions of the cerebral cortex are associated with sensory input or with motor output. You then discovered that the electrical activity of the brain can be recorded in the form of electroencephalograms (EEGs). Finally, you read about the role of the brain in sleep and memory.

In the last part of this section you read about the spinal cord. The spinal cord acts as a communications link between the brain and the peripheral nervous system. It also regulates reflexes.

### Applying Definitions: Building Vocabulary Skills

Label the following parts of the brain on the accompanying illustration: cerebellum, cerebrum, hypothalamus, medulla oblongata, midbrain, pons, and thalamus.

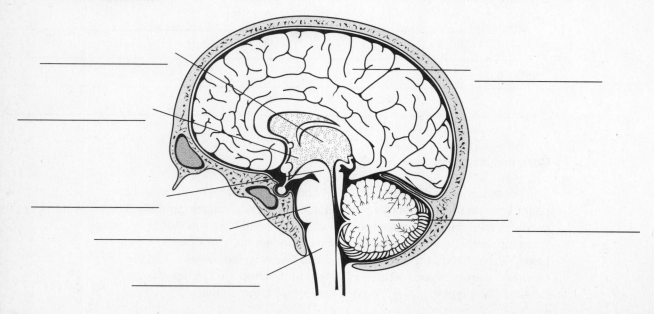

© Prentice-Hall, Inc.

### How's Your Brain Functioning?: Using the Main Ideas

Identify the part of the central nervous system that controls each of the following activities. Be as specific as possible. For example, if the answer is *brain,* name the particular part of the brain that controls the activities, such as *cerebrum* or *medulla oblongata.*

1. Dancing: _____

2. Regulating reflexes: _____

3. Linking the cerebral cortex and the cerebellum: _____

4. Making decisions: _____

5. Walking: _____

6. Controlling heartbeat: _____

7. Connecting the brain and spinal cord: _____

8. Coordinating muscle: _____

9. Answering this question: _____

10. Regulating body temperature: _____

11. Sneezing: _____

12. Linking the brain and peripheral nervous system: _____

13. Controlling voluntary actions: _____

14. Storing memories: _____

15. Falling asleep: _____

### Concept Mapping

The construction of and theory behind concept mapping are discussed on pages vii–ix in the front of this Study Guide. Read those pages carefully. Then consider the concepts presented in Section 37–3 and how you would organize them into a concept map. Now look at the concept map for Chapter 37 on page 362. Notice that the concept map has been started for you. Add the key facts and concepts you feel are important for Section 37–3. When you have finished the chapter, you will have a completed concept map.

| Section 37–4 | The Peripheral Nervous System | (pages 825–826) |

## SECTION REVIEW

In this section you were introduced to the peripheral nervous system, which connects the central nervous system with the rest of the body. The peripheral nervous system is divided into the sensory and motor divisions. The sensory division provides the brain and spinal cord with information from the sense organs. The motor division carries information from the central nervous system to the muscles and glands.

The motor division is further divided into the somatic nervous system and the autonomic nervous system. The somatic nervous system controls activities that are under conscious control. Activities such as dancing, running, and playing the piano are examples of activities carried out by muscles controlled by the somatic nervous system. But because many of the nerves within the somatic nervous

system are part of reflexes—automatic responses to stimuli—they can cause the muscles to act involuntarily.

As its name suggests, the autonomic nervous system controls muscles that are involuntary or "automatic." Examples of activities that are involuntary and under the control of the autonomic nervous system are the contractions of the heart muscle and the contractions of muscles in the digestive system. The autonomic nervous system is divided into the sympathetic nervous system and the parasympathetic nervous system. These two subdivisions of the autonomic nervous system have opposite effects on the organs they control. Because of these opposite effects, the sympathetic and parasympathetic nervous systems can turn organ functions "on" and "off."

### Applying Definitions: Building Vocabulary Skills

Complete the equations to show the relationships among the parts of the peripheral nervous system.

1. Peripheral nervous system = _____

   + _____

2. Sympathetic nervous system + _____

   = Autonomic nervous system

3. Cranial nerves + Spinal nerves + Ganglia = _____

   _____

4. Somatic nervous system + Autonomic nervous system = _____

   _____

5. Nose + Ears + Eyes + Taste buds + Skin = _____

   _____

■ **Applying Concepts: The Reflex Arc**

**1.** Label the diagram of the reflex arc using the following terms: Sensory neuron (receptor), Motor neuron, Spinal cord, Muscle (effector), Heat receptor, Interneuron.

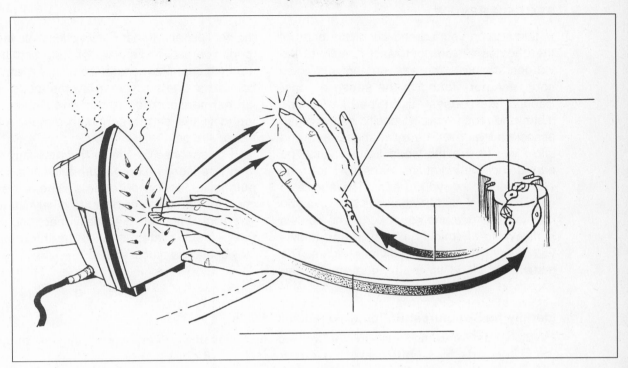

**2.** Why is it important that the body have a shortcut method of responding to harmful stimuli? _____

_____

_____

_____

■ **Concept Mapping**

The construction of and theory behind concept mapping are discussed on pages vii–ix in the front of this Study Guide. Read those pages carefully. Then consider the concepts presented in Section 37–4 and how you would organize them into a concept map. Now look at the concept map for Chapter 37 on page 362. Notice that the concept map has been started for you. Add the key facts and concepts you feel are important for Section 37–4. When you have finished the chapter, you will have a completed concept map.

| Section 37–5 | **The Senses** | *(pages 827–831)* |

## SECTION REVIEW

In this section you learned about the parts of the nervous system that react directly to the various stimuli in the environment: the sensory receptors found in the sense organs. Each of the senses—sight, hearing, smell, taste, and touch—has a specific sense organ associated with it. The eyes sense light and allow you to see the world around you. The ears detect the vibrations in the air known as sounds and allow you to hear. The ears also contain special fluid-filled canals that help you maintain balance and equilibrium. The nose is sensitive to chemicals in the air and allows you to smell thousands of different odors. The taste buds found on the tongue, roof of the mouth, lips, and throat detect chemicals in the foods you eat. The senses of taste and smell are both chemical senses, and you depend on both senses to enjoy your favorite foods. The largest sense organ is the skin. Receptors found in the skin respond to temperature, touch, and pain.

In response to stimuli from the environment, sensory receptors produce nerve impulses, or action potentials. These nerve impulses can be transmitted to other neurons. The nerve impulses eventually reach the central nervous system, where the information they contain is processed.

### Identifying Structures: Building Vocabulary Skills

**1.** Label the following structures on the accompanying diagram of the eye: aqueous humor, cornea, iris, lens, optic nerve, pupil, retina, sclera, vitreous humor.

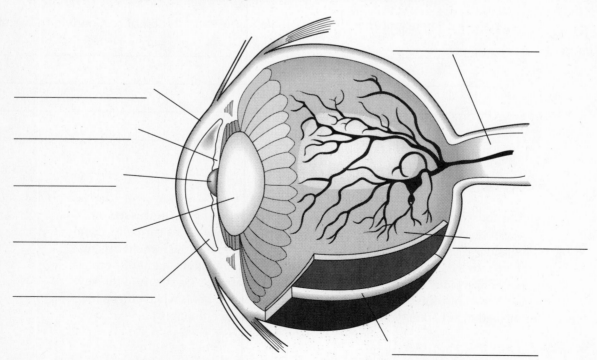

**2.** In the space provided, write the name of the structure that best fits each description. (*Hint:* You labeled all these structures on the above diagram.)

**a.** Muscles attached to this part of the eye allow you to focus on objects: _____

**b.** Regulates the amount of light that enters the eye: _____

**c.** Contains rod and cone cells that convert light energy into nerve impulses:

_____

**d.** Tough white tissue that maintains the shape of the eye: _____

**e.** Transparent front of the eye through which light passes: _____

**f.** Small opening through which light enters the eye: _____

**g.** Fluid located in a small chamber between the cornea and the iris:

_____

### Applying Concepts: Using the Main Ideas

Fill in the blanks in the following paragraph. You may wish to refer to the diagram of the ear on page 830 in your textbook to help you correctly complete this exercise.

As your favorite song begins playing on the radio, the sound coming out of the speakers

causes the air to vibrate. Your _____ collects these vibrations and

channels them into the _____. The tiny hairs and wax in your

_____ prevent foreign objects, which might interfere with your

hearing, from entering your ears. The sound vibrations strike your

_____ and are transmitted to the bones of the middle ear: the

_____, _____, and _____.

The _____ vibrates against the thin membrane covering your

_____, which then transmits the vibrations to the fluid-filled

_____. As the fluid in your _____ vibrates,

tiny hairs are pushed back and forth, providing stimulation that is turned into nerve

impulses. These nerve impulses are carried to the brain by the _____,

and you are able to enjoy your favorite song. When the song is over, you decide to

change the radio station. When you get up off the floor, the fluid and otoliths in the

_____ of your middle ear send impulses to your brain that help

you maintain your balance.

### Relating Concepts: Using the Main Ideas

1. Identify the sense organ or organs that would respond to the following stimuli.

   **a.** Temperature: _____

   **b.** Sound: _____

   **c.** Chemicals: _____

   **d.** Body movement: _____

   **e.** Light: _____

   **f.** Pressure: _____

   **g.** Pain: _____

2. Why do you think that the senses of taste and smell are related?

   _____

   _____

   _____

   _____

   _____

   _____

   _____

   _____

### Concept Mapping

The construction of and theory behind concept mapping are discussed on pages vii–ix in the front of this Study Guide. Read those pages carefully. Then consider the concepts presented in Section 37–5 and how you would organize them into a concept map. Now look at the concept map for Chapter 37 on page 362. Notice that the concept map has been started for you. Add the key facts and concepts you feel are important for Section 37–5. When you have finished the chapter, you will have a completed concept map.

**Using the Writing Process**                                   *Chapter 37*

Use your writing skills and imagination to respond to the following writing assignment. You will probably need an additional sheet of paper to complete your response.

The inhabitants of the gas-giant planet Zephyr live in the middle portion of the atmosphere. Although Zephyrians are intelligent, they are quite different from humans. They look like wisps of smoke or clouds and lack a fixed form. Their sensory receptors, which are distributed evenly throughout their bodies, perceive the presence of chemicals and changes in pressure. Their bodies basically respond to stimuli by condensing or by becoming more diffuse.

Now imagine that you are a Zephyrian scientist. You have managed to transfer your consciousness into a human body on Earth. As well as being able to perceive your surroundings through your host's senses, you are also able to observe the workings of the human nervous system. From your alien perspective, describe what it is like to have a human nervous system.

*Hints:* What senses are different in humans and Zephyrians? How do the responses of human and Zephyrian bodies differ? What advantages and disadvantages would a Zephyrian perceive in the way a human senses the environment?

You may wish to select someone you know—yourself, a friend, or a classmate, for example—as a host for the Zephyrian.

_____

_____

_____

_____

_____

_____

_____

_____

_____

_____

_____

_____

_____

_____

## Concept Mapping *Chapter 37*

The concept map below has been started for you. Add the key facts and concepts for each section of the chapter to this partial concept map. When you are done, you will have a concept map for the entire chapter.

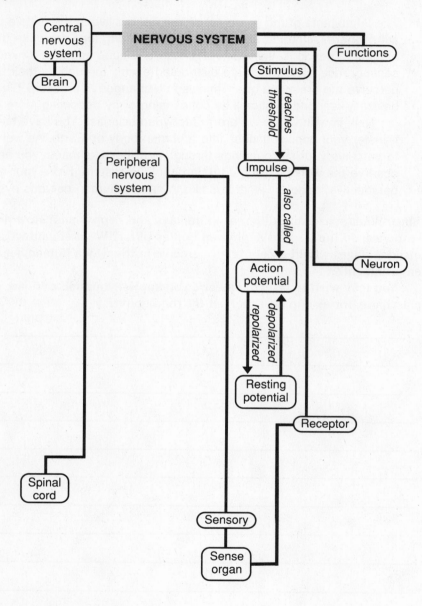

S T U D Y  
G U I D E

CHAPTER **38**

*Skeletal, Muscular, and Integumentary Systems*

| Section 38–1 | **The Skeletal System** | *(pages 837–842)* |

### SECTION REVIEW

In this section you learned about the bones, cartilage, tendons, and ligaments that make up the skeletal system. The bones of the skeletal system work together to support and shape the body, protect vital internal organs, provide a place for muscles to attach to allow movement, store minerals, and produce blood cells.

You also learned about the structure of the bones in the body. The tough membrane that surrounds each bone is the periosteum. Beneath the periosteum is hard, compact bone. Running through compact bone is a network of Haversian canals that contain blood vessels and nerves. Inside the layer of compact bone is spongy bone. Spongy bone, despite its name, is quite strong and adds strength to bones without adding too much mass. Embedded in compact and spongy bone are osteocytes that can either deposit calcium salts in bone or absorb them again. Osteocytes are responsible for the growth of the bones and the changes in their shapes.

Cavities within bones contain bone marrow, a special soft tissue that produces red and white blood cells.

You also learned that bones form when soft cartilage is replaced by harder bone during the process of ossification. During ossification, the minerals calcium and phosphorus are deposited near the center of the bone. Bone tissue forms as osteocytes secrete minerals that replace the soft cartilage with bone cells. The process of ossification begins up to 7 months before birth and continues for 18 to 20 years.

You also learned that joints are places where bones meet. Joints allow movement without damaging the bones. The joints in the body are classified according to the amount of movement they allow. Immovable joints allow no movement between bones; slightly movable joints allow some movement; and the six types of freely movable joints allow a wide range of movements.

### Applying Concepts: Building Vocabulary Skills

**1.** Label the parts of the bone shown in the diagram below.

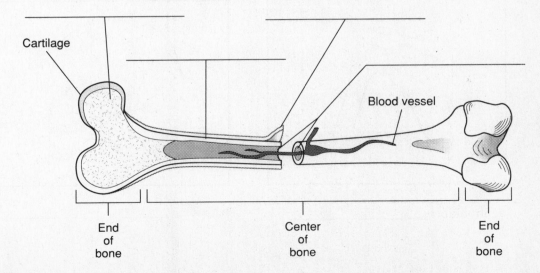

Cartilage

Blood vessel

End of bone

Center of bone

End of bone

2. Identify the part of the skeletal system described by each of the following phrases.

    **a.** Flexible connective tissue that is replaced by bone during ossification:

    _____

    **b.** Tough connective tissue that holds bones together: _____

    **c.** Type of connective tissue that makes up the skeletal system: _____

    **d.** Forms during the process of ossification: _____

    **e.** Found in parts of the body where flexibility is needed (nose, ears, and sternum):

    _____

## Structures and Functions: Relating Concepts

Using the appropriate letter(s), match the part(s) of the bone with the function of the skeletal system described in each numbered phrase.

    **A.** Periosteum    **C.** Compact bone    **E.** Joints
    **B.** Spongy bone    **D.** Bone marrow    **F.** Osteocytes

    **1.** Provide protection for vital internal organs: _____

    **2.** Produces red and white blood cells:_____

    **3.** Supports and shapes the body: _____

    **4.** Store minerals: _____

    **5.** Allow movement: _____

## Identifying Joints: Using the Main Ideas

Study each example of a joint in the accompanying diagram. In the space provided, identify the type of joint.

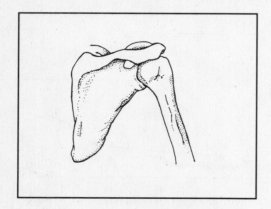
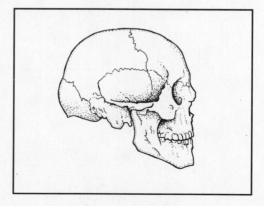

    **a.** Shoulder: _____    **b.** Skull: _____

                                                              

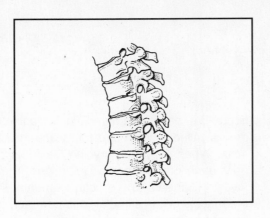

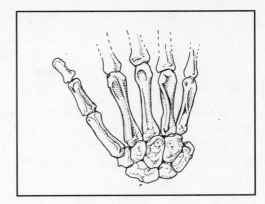

**c.** Backbone: _____

**d.** Base of thumb: _____

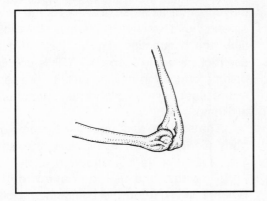

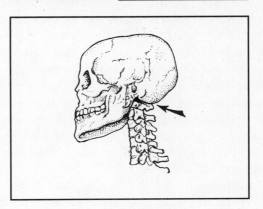

**e.** Elbow: _____

**f.** Top of neck: _____

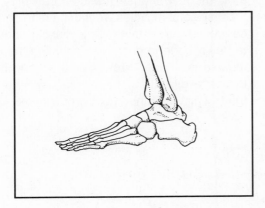

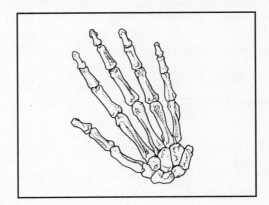

**g.** Ankle: _____

**h.** Base of fingers: _____

### Concept Mapping

The construction of and theory behind concept mapping are discussed on pages vii–ix in the front of this Study Guide. Read those pages carefully. Then consider the concepts presented in Section 38–1 and how you would organize them into a concept map. Now look at the concept map for Chapter 38 on page 372. Notice that the concept map has been started for you. Add the key facts and concepts you feel are important for Section 38–1. When you have finished the chapter, you will have a completed concept map.

**Section 38–2**   **The Muscular System**                          *(pages 842–846)*

### SECTION REVIEW

In this section you learned that the muscular system allows the body to move. Movement is produced when specialized muscle tissue contracts, or shortens, when it is stimulated by a nerve impulse.

You also learned that there are three different types of muscle tissue, or muscles, within the body: skeletal, smooth, and cardiac. Each type of muscle tissue has its own structure and function. Skeletal muscles, as their name indicates, are generally attached to bones and are responsible for moving the bones. These muscles allow voluntary movements because they are under the conscious control of the central nervous system. Smooth muscles are located in many internal organs and in the walls of blood vessels. These muscles are not under conscious control and are responsible for involuntary actions such as moving food through the digestive system, controlling blood flow through blood vessels, and increasing or decreasing the size of the pupils of the eyes. Cardiac muscle is found only in the heart. Like smooth muscle, cardiac muscle is not under voluntary control: You do not have to think about making your heart beat.

You also learned that the sliding filament theory explains how muscles contract. The fibers that make up a muscle cell are thick filaments called myosin and thin filaments called actin. When actin and myosin filaments come near each other, cross-bridges form from the myosin filaments to the actin filaments. When the muscle cell is stimulated to contract, these cross-bridges move and pull the filaments past each other. When each cross-bridge has moved as far as it can, it releases the actin filament and reattaches at another place to continue the contraction. The process of muscle contraction requires energy. ATP is constantly being produced in muscle cells to provide the energy necessary to contract muscles and allow movement.

The skeletal and muscular systems of the body work together to produce movement. Skeletal muscles are attached to bones by tough connective tissues called tendons. When a skeletal muscle contracts, it pulls on the bone to which it is attached and produces movement. Because muscles can only pull, most skeletal muscles work in pairs to produce controlled movements. One muscle contracts to pull a bone in one direction while the other muscle is relaxed. To move the bone in the opposite direction, the relaxed muscle contracts while the other muscle now relaxes.

### Identifying Functions: Applying Vocabulary Skills

Identify the type or types of muscle tissue described in each of the following items.

**1.** Attached to bones: _____

**2.** Found only in the heart: _____

**3.** Made up of bundles of muscle fibers: _____

**4.** Found in internal organs: _____

**5.** Responsible for voluntary movement: _____

**6.** Composed of actin and myosin filaments: _____

**7.** Not under voluntary control: _____

**8.** Can contract without direct stimulation from the nervous system: _____

**9.** Moves food through the digestive system: _____

**10.** Allows you to walk: _____

### Muscle Contraction: Relating Concepts

**1.** Label the actin, myosin, and cross-bridges in the accompanying diagram.

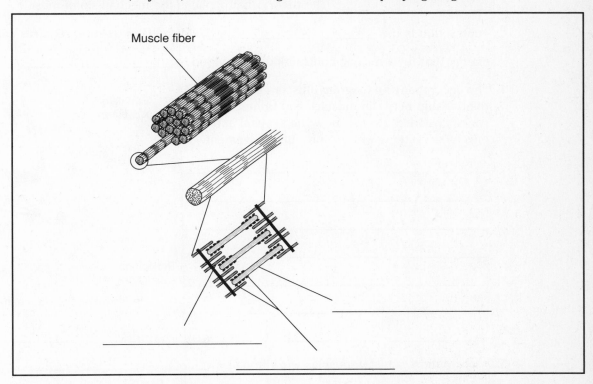

Muscle fiber

**2.** The accompanying diagrams show a contracted and a relaxed muscle. Identify each diagram by properly labeling it "contracted" or "relaxed."

**3.** The following paragraph describes the contraction of a muscle. Complete the description by providing the correct word or words to finish each sentence.

Skeletal muscles are made of filaments called _____.

When these filaments get close to one another, small projections in each myosin filament

_____. When the muscle is stimulated

by a _____ to contract, the cross-bridges

_____. When a
cross-bridge has moved as far as it can, the actin filament is released and returns to its
original position. The cross-bridge then reattaches to the _____
at another place and _____. As thousands of actin and
myosin filaments form cross-bridges and move, the entire muscle cell
_____. The theory that explains the mechanism of muscle
contraction is the _____. Muscle contraction requires energy. The
energy that fuels muscle contraction is provided by_____.

4. The accompanying diagram illustrates the
   relationship between muscles and bones. In the
   space provided, explain how the skeletal and
   muscular systems work to produce movement.

   Bundle of
   muscle fibers

   Tendon

   Periosteum
   of bone

   Bone marrow

   _____

   _____

   _____

   _____

   _____

   _____

   _____

5. Why do most skeletal muscles work in pairs? _____

   _____

   _____

   _____

   _____

   _____

### Concept Mapping

The construction of and theory behind concept mapping are discussed on
pages vii–ix in the front of this Study Guide. Read those pages carefully. Then
consider the concepts presented in Section 38–2 and how you would organize
them into a concept map. Now look at the concept map for Chapter 38 on
page 372. Notice that the concept map has been started for you. Add the key
facts and concepts you feel are important for Section 38–2. When you have
finished the chapter, you will have a completed concept map.

## SECTION REVIEW

In this section you learned about the organs of the integumentary system and how they protect the body. The skin, hair, and nails of the integumentary system protect the body from infection, injury, and ultraviolet radiation; help maintain a constant body temperature; and remove waste products from the body. The integumentary system also functions as a sense organ by collecting information from the environment and sending it to the nervous system.

The largest organ in the body and the main organ of the integumentary system is the skin. Skin is made of two layers. The epidermis is the outer layer and is made of cells that divide rapidly to constantly produce new cells. Older cells in the epidermis produce the protein keratin, which forms the basic struc-

ture of hair and nails. The epidermis also contains cells that produce melanin, the pigment that gives the skin its color and offers protection from the sun's ultraviolet radiation. The epidermis provides the waterproof cover for the skin and serves as a barrier to disease-causing organisms. The dermis, or inner layer of skin, contains blood vessels and fat, which help regulate body temperature. The sweat glands in the dermis also help maintain body temperature and help rid the body of waste products by secreting sweat. Sebaceous, or oil, glands keep the surface of the skin flexible and waterproof. The nerve endings in the dermis function as sense organs and constantly gather information from the environment for the nervous system.

### Structures and Functions: Relating Concepts

Identify the organ or organs of the integumentary system that carry out each function listed below.

**1.** Protect against infection and injury: _____

_____

**2.** Regulate body temperature: _____

_____

**3.** Removes waste products: _____

_____

**4.** Protects from ultraviolet radiation: _____

_____

**5.** Allows for sensory information: _____

_____

### Applying Definitions: Building Vocabulary Skills

**1.** The Latin word *integument* means to cover. How does the Latin definition of integument relate to the function of the integumentary system?

_____

_____

_____

_____

**2.** Identify the parts of the skin shown in the accompanying diagram.

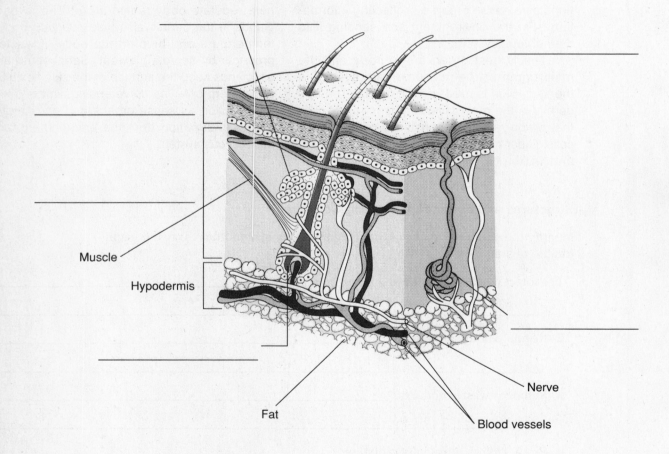

Muscle

Hypodermis

Fat

Nerve

Blood vessels

### Concept Mapping

The construction of and theory behind concept mapping are discussed on pages vii–ix in the front of this Study Guide. Read those pages carefully. Then consider the concepts presented in Section 38–3 and how you would organize them into a concept map. Now look at the concept map for Chapter 38 on page 372. Notice that the concept map has been started for you. Add the key facts and concepts you feel are important for Section 38–3. When you have finished the chapter, you will have a completed concept map.

## Using the Writing Process

Use your writing skills and imagination to respond to the following writing assignment. You will probably need an additional sheet of paper to complete your response.

We have many different "appreciation days" in our society. Mother's Day and Father's Day are two examples. Suppose that it has been decided that each body system should have its own day of recognition. You have been selected to design an appreciation day for the skeletal and muscular systems. Write an agenda for this day describing each activity on the program and explaining why you have chosen it.

_____

_____

_____

_____

_____

_____

_____

_____

_____

_____

_____

_____

_____

_____

_____

_____

_____

_____

_____

## Concept Mapping

The concept map below has been started for you. Add the key facts and concepts for each section of the chapter to this partial concept map. When

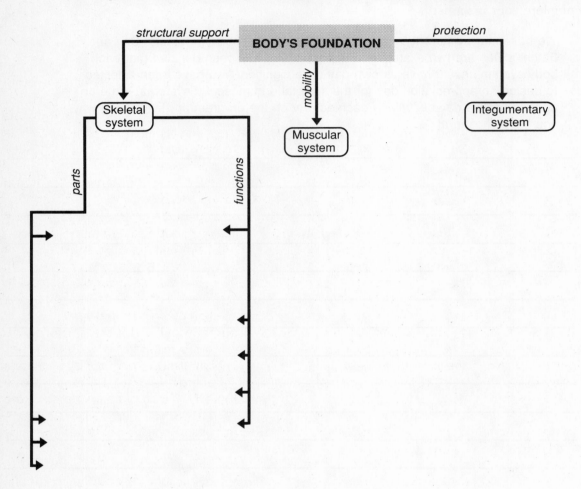

S T U D Y
G U I D E

CHAPTER
Nutrition and Digestion **39**

| Section 39–1 | **Food and Nutrition** | *(pages 855–866)* |

## SECTION REVIEW

In this section you learned that besides being enjoyable, eating is important for the proper functioning of your body. When you eat a variety of foods from the four basic food groups, you are supplying your body with the nutrients it needs in order to function properly. The following are the four basic food groups: meat, fish, and beans; milk and milk products; fruits and vegetables; and bread, rice, and cereals. These four food groups provide your body with the essential nutrients it needs: water, minerals, carbohydrates, fats, proteins, and vitamins.

You also learned that the nutrients found in the four basic food groups provide your body with energy, material for growth and repair, and the substances necessary for proper functioning. Water is a nutrient that you cannot live without. Water is the solvent in which food, enzymes, and waste materials are dissolved, and it is used to help regulate your body's temperature. Water is readily available in the liquids you drink and in the foods you eat and is a product of cellular respiration.

Minerals are substances that your body needs to function properly. They are available in vegetables, meats, table salt, and seafood.

Carbohydrates and fats are nutrients that supply your body with sources of energy. Carbohydrates may be simple sugars that provide a quick source of energy or complex starches that provide energy only after they have been broken down by the digestive system. Another important carbohydrate is cellulose, which is found in the cell walls of plants. You cannot digest cellulose, but it provides roughage, which stimulates the muscles of the digestive system and enables your digestive system to work more efficiently.

In addition to storing energy for future use, fats provide protection for vital organs, assist in temperature regulation, and keep the skin from drying out. Fats are therefore important in your diet and are obtained in many of the cooking oils we use to prepare foods. But too much fat in a diet can create problems with blood circulation. Fat deposits in arteries can restrict the flow of blood to vital organs.

Proteins provide the body with amino acids, which are materials for the growth and repair of body tissues, and serve as another source of energy. Of the 20 different amino acids required, 12 can be manufactured by the body. The other 8, which are called the essential amino acids, must be obtained from the diet.

Small amounts of vitamins are needed to put together the body's building materials. There are two groups of vitamins: fat-soluble vitamins, which can be stored in the body for future use, and water-soluble vitamins, which should be part of your diet every day because they cannot be stored by the body. By selecting foods from all four food groups, you can be sure to get all the nutrients your body requires.

**Identifying Nutrients: Using the Main Ideas**

Complete the chart that follows.

| Nutrient | Source | Function(s) |
|---|---|---|
| Water | | |
| | | 1. Protection for vital organs<br>2. Energy source<br>3. Help in temperature regulation<br>4. Keep skin from drying out |
| | Meat, fish, beans, dairy products | |
| | | 1. Provide roughage to aid in digestion<br>2. Energy source |
| Minerals | All types of foods | |
| | | 1. Help regulate body processes<br>2. Help enzymes start reactions |

### ▨ The Four Food Groups: Relating Concepts

1. Match each food item to the letter of the food group to which it belongs.

**A.** Meat, fish, and beans          **C.** Fruits and vegetables
**B.** Milk and milk products        **D.** Bread, rice, and cereals

Steak: _____

Oranges: _____

Macaroni: _____

Peanuts: _____

Corn: _____

Oatmeal: _____

Eggs: _____

Tortillas: _____

2. Using the key below, identify the foods that are high in the following nutrients.

P—Proteins          F—Fats
C—Carbohydrates     V—Vitamins

Fish: _____

Carrots: _____

Oranges: _____

Candy: _____

Chicken: _____

Bread: _____

Beans: _____

Butter: _____

Cooking oil: _____

Broccoli: _____

### A Balanced Diet: Applying Definitions

List the foods you ate at your last three meals in the appropriate place in
the chart. Then answer the questions that follow.

| Meat, Fish and Beans | Milk and Milk Products | Fruits and Vegetables | Bread, Rice and Cereals |
|---|---|---|---|
|  |  |  |  |
|  |  |  |  |
|  |  |  |  |

**1.** What is a balanced diet? _____

_____

_____

_____

_____

_____

**2.** Based on the information in the chart, are you eating a balanced diet? Explain your

answer. _____

_____

_____

_____

_____

_____

### Concept Mapping

The construction of and theory behind concept mapping are discussed on
pages vii–ix in the front of this Study Guide. Read those pages carefully. Then
consider the concepts presented in Section 39–1 and how you would organize
them into a concept map. Now look at the concept map for Chapter 39 on
page 382. Notice that the concept map has been started for you. Add the key
facts and concepts you feel are important for Section 39–1. When you have
finished the chapter, you will have a completed concept map.

## Section 39-2    The Process of Digestion                    *(page 867–875)*

### SECTION REVIEW

In this section you learned that the foods you eat must be broken down into simpler molecules so that the nutrients can be absorbed and used by the cells of the body. Digestion is the process that breaks down food into these simpler molecules. The process of digestion can be divided into three parts. First, food is broken down into smaller pieces during mechanical digestion. Next, these smaller pieces of food are broken down into simpler molecules during chemical digestion. Finally, these simpler nutrient molecules are absorbed from the digestive tract into the bloodstream, where they are transported to individual cells.

You also learned about the organs and glands of the digestive system and how they contribute to the process of digestion. Digestion begins in the mouth when your teeth chew the food you eat into smaller pieces and saliva from your salivary glands begins the chemical digestion of starches into simpler sugars. When you swallow, food passes through the throat, or pharynx, and enters the esophagus.

The esophagus is a long muscular tube that moves the food toward the stomach by muscular contractions called peristalsis. When food enters the stomach, it is mixed with gastric fluids by mechanical churning. The gastric fluids secreted by glands in the stomach begin the digestion of proteins. After several hours in the stomach, partly digested food, called chyme, enters the duodenum of the small intestine. With the help of bile from the liver and pancreatic fluid from the pancreas, the chemical digestion of proteins, fats, and carbohydrates is completed in the duodenum. As digested food continues into the jejunum and ileum of the small intestine, nutrient molecules are absorbed into the bloodstream through fingerlike projections called villi. Food material that cannot be digested is passed into the large intestine, or colon. In the large intestine, water is removed and undigested material, with the help of bacteria, becomes solid waste, or feces, that will be eliminated from the body through the anus.

### Applying Concepts: Building Vocabulary Skills

1. Replace the underlined phrase with the correct vocabulary term.

   a. The nutrients in the food you eat are made available to the cells of your body when they are broken down into simpler molecules. _____

   b. The process of breaking food into smaller pieces increases the amount of food exposed to digestive chemicals. _____

   c. Nutrient molecules that can be used by your body's cells are produced during the process of breaking complex molecules into simpler molecules.

      _____

   d. In the small intestine, digestion is completed and the end products of digestion are transferred from the small intestine to the bloodstream.

      _____

   e. A series of muscular contractions moves food through the digestive system.

      _____

**2.** Complete the equation:

Mechanical digestion + _____ + _____ = Digestion

## The Digestive System: Identifying Structures

Label the organs of the digestive system in the diagram below.

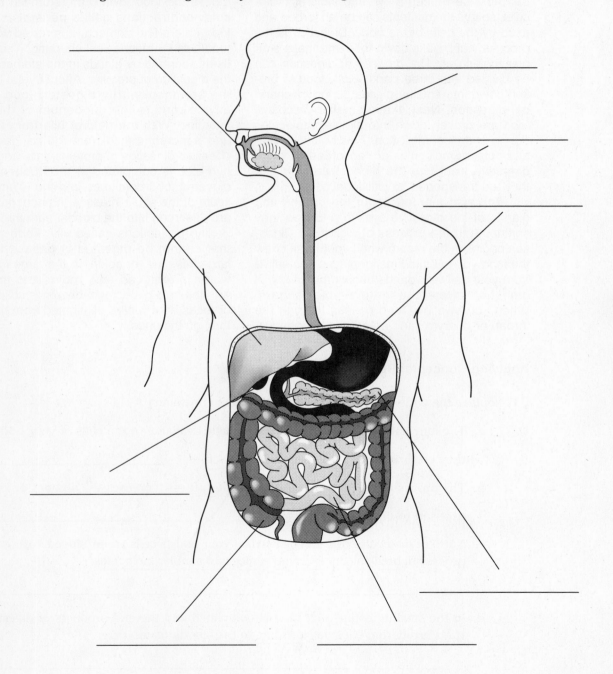

**Structures and Functions: Relating Concepts**

Complete the following chart.

| Organ | Type of Digestion | Digestive Secretion or Enzyme | Function |
|---|---|---|---|
| Mouth | | | |
| | Mechanical | None | |
| | | Pepsin, hydrochloric acid, and mucus | |
| Small intestine | | | |
| Liver | None | | |
| | None | | Aids in digestion of proteins and carbohydrates |
| Large intestine | None | None | |

### Sequencing Events: Applying the Main Ideas

Using the numbers 1 through 8, place the digestive events described below in the correct sequence.

_____ Chyme moves into the duodenum, where bile from the liver, pancreatic fluid from the pancreas, and enzymes from the intestine itself complete the chemical digestion of proteins, fats, and carbohydrates.

_____ Peristalsis moves food down the esophagus and into the stomach.

_____ Saliva begins the digestion of starch.

_____ As food passes through the jejunum and ileum, nutrients are absorbed into the bloodstream through the villi.

_____ Churning of the stomach mixes food with gastric fluids.

_____ Water is absorbed in the colon and solid waste material is prepared for elimination.

_____ Chewed food passes through the pharynx and into the esophagus.

_____ Each bite of food is chewed and mixed with saliva.

### Food and Digestion: Relating Main Ideas

Pretend that you have just finished eating a steak, a baked potato full of butter, hot rolls with butter, and a green salad with oil and vinegar dressing. Identify the digestive organ in which the chemical digestion of each of these foods would begin. (*Hint:* think about the nutrients found in each of these food items.)

**1.** Steak: _____

**2.** Potato: _____

**3.** Butter: _____

**4.** Rolls: _____

**5.** Lettuce: _____

**6.** Salad dressing: _____

### Concept Mapping

The construction of and theory behind concept mapping are discussed on pages vii–ix in the front of this Study Guide. Read those pages carefully. Then consider the concepts presented in Section 39–2 and how you would organize them into a concept map. Now look at the concept map for Chapter 39 on page 382. Notice that the concept map has been started for you. Add the key facts and concepts you feel are important for Section 39–2. When you have finished the chapter, you will have a completed concept map.

## Using the Writing Process

Use your writing skills and imagination to respond to the following writing assignment. You will probably need an additional sheet of paper to complete your response.

People often use satire to make a statement. Satire is a literary technique that uses irony and exaggeration to make a point. Imagine you are a nutritionist who is totally frustrated by some of the strange diets advocated in a variety of diet books. Combat this problem by writing a satire on diet books.

_____

_____

_____

_____

_____

_____

_____

_____

_____

_____

_____

_____

_____

_____

_____

_____

_____

_____

_____

## Concept Mapping *Chapter 39*

The concept map below has been started for you. Add the key facts and concepts for each section of the chapter to this partial concept map. When you are done, you will have a concept map for the entire chapter.

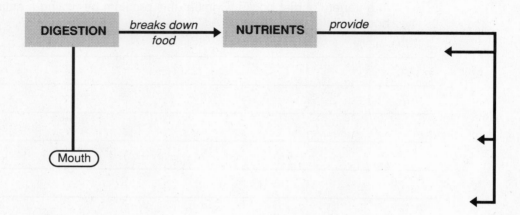

S T U D Y
G U I D E

| Section 40–1 | The Importance of Respiration | *(pages 881–882)* |

### SECTION REVIEW

In this section you learned that the process of respiration allows living organisms to exchange oxygen and carbon dioxide with their environment. Each time you breathe, the respiratory system takes in oxygen and gets rid of carbon dioxide, a waste material. This type of respiration is called external respiration because you exchange gases with your environment. Inside the cells of the body, another type of respiration is also taking place: internal respiration. During internal respiration, the mitochondria in the cells use the oxygen obtained during external respiration to produce large amounts of energy in the form of ATP. Carbon dioxide is produced as a waste product during inter-

nal respiration and is then expelled from the body during external respiration.

You also learned that single-celled organisms do not need respiratory systems because gas exchange with the environment takes place across the cell membrane. Multicellular organisms have many cells that do not come into contact with the environment. But because these cells still require oxygen and produce carbon dioxide, they require a special system to carry out respiration efficiently: the respiratory system. Without an efficient respiratory system, multicellular organisms could not survive.

### Applying Definitions: Building Vocabulary Skills

1. *Internal* means inside; *external* means outside. Why are these terms used to describe the two types of respiration that occurs in the body? _____

    _____

    _____

    _____

    _____

    _____

2. External respiration involves an organism's respiratory system. What is the function of the respiratory system? _____

    _____

    _____

    _____

    _____

### Relating Concepts: Using the Main Ideas

1. In a large multicellular organism, could internal respiration take place without external

   respiration? Explain your answer. _____

   _____

   _____

   _____

   _____

   _____

   _____

   _____

2. Why is a respiratory system not necessary in a single-celled organism?

   _____

   _____

   _____

   _____

   _____

   _____

   _____

   _____

### Concept Mapping

The construction of and theory behind concept mapping are discussed on
pages vii–ix in the front of this Study Guide. Read those pages carefully. Then
consider the concepts presented in Section 40–1 and how you would organize
them into a concept map. Now look at the concept map for Chapter 40 on
page 392. Notice that the concept map has been started for you. Add the key
facts and concepts you feel are important for Section 40–1. When you have
finished the chapter, you will have a completed concept map.

## Section 40-2  The Human Respiratory System          *(pages 882–889)*

### SECTION REVIEW

In this section you learned about the respiratory system. The respiratory system is composed of the nose, larynx, pharynx, trachea, bronchi, and lungs. The respiratory system can best be described as a series of passageways that not only direct the air that is inhaled to the lungs, where gas exchange takes place, but also constantly clean and filter out impurities that might interfere with the functioning of this important system.

In order to air into the respiratory system, you inhale. When you inhale, the diaphragm, a large muscle located at the bottom of the rib cage, contracts and moves down. When the diaphragm moves down, the volume of the chest cavity increases. With an increase in volume, there is also a decrease in pressure. The air outside the body is still at atmospheric pressure; so to equalize the pressure inside and outside the body, air rushes into the respiratory system through the nose, pharynx, and trachea.

Once inside the nose, air from the environment is filtered by tiny hairs called cilia and mucous secretions that trap particles of dirt, germs, and other foreign objects. Because the air you breathe is often very dry, moisture is added as it passes through the nose and into the pharynx. Air then passes into the trachea for its trip into the chest cavity. Once again, mucus and cilia work together to filter out any particles still in the air. The trachea divides into two branches called bronchi that lead directly into the lungs. The lungs are the main organs of the respiratory system, and it is in the lungs that gas exchange actually takes place. The bronchi divide into smaller and smaller passageways called bronchioles that eventually lead into the hollow air sacs called alveoli. The alveoli are thin membranes that are covered with tiny blood vessels called capillaries. In the alveoli, the oxygen and carbon dioxide are exchanged between the air and the bloodstream.

The concentration of oxygen in the air that is inhaled is greater than the concentration of oxygen in the bloodstream. Because of the difference in concentrations of oxygen, oxygen diffuses across the thin membrane of the alveoli into the bloodstream, where it is carried to the cells in the body. A special oxygen-carrying molecule, hemoglobin, found in red blood cells, enables the body to carry the large amount of oxygen needed by the cells of the body. At the same time, the concentration of the waste gas carbon dioxide is greater in the bloodstream than in the air that was inhaled. Because of the difference in concentrations of carbon dioxide, carbon dioxide diffuses from the bloodstream into the air in the alveoli so that it can be exhaled from the body.

When you exhale, you force the air that now contains the waste gas carbon dioxide out of the respiratory system. When you exhale, the diaphragm relaxes and moves back into its position at the base of the rib cage. This action decreases the volume of the chest cavity and increases the pressure inside the chest. The increased pressure inside the chest cavity causes the lungs to decrease in size and to squeeze the air containing carbon dioxide out of the alveoli, the lungs, and the body.

As the exhaled air reverses its trip through the passages of the respiratory system, it rushes past the vocal cords in the larynx. When muscles cause the vocal cords to contract, the air passing between them vibrates and produces sound. The respiratory system allows the body to efficiently exchange vital gases with the environment and provides a mechanism for transferring gases into the bloodstream so that each cell of the body receives the material it needs to carry out its function.

## The Respiratory System: Applying Vocabulary Skills

1. Label the organs of the respiratory system in the diagram below.

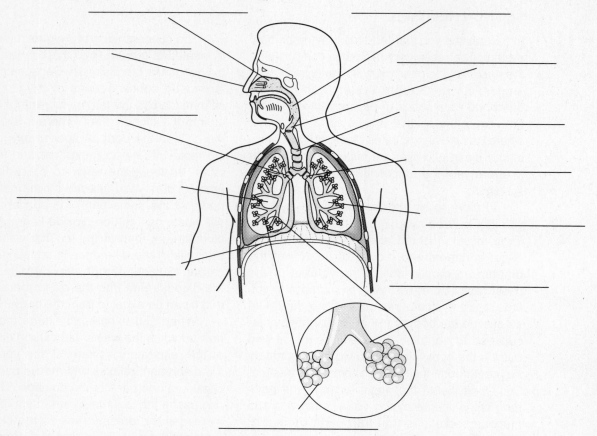

2. Identify the organ or organs of the respiratory system that perform the following functions.

a. Passageway for air from the environment into the respiratory system:

_____

b. Clean and filter the air from the environment: _____

c. Contain the alveoli: _____

d. Specific location of gas exchange: _____

e. Moisture is added to the air: _____

f. Contains the vocal cords: _____

g. Muscle that contracts and relaxes to cause inhalation and exhalation:

_____

h. Place where exhaled air leaves the respiratory system:

_____

### Interpreting Illustrations: The Mechanics of Breathing

1. In the space to the right of each diagram, identify and describe the process illustrated by each diagram.

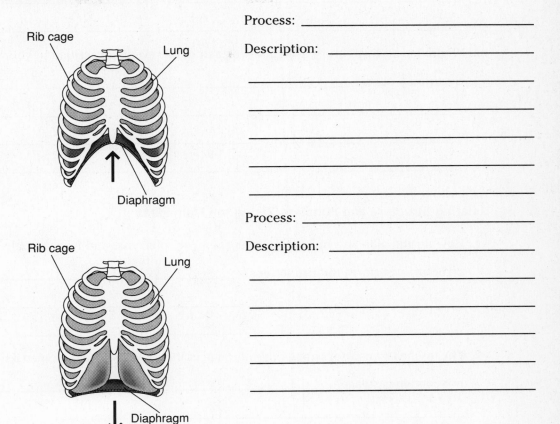

Rib cage

Lung

Diaphragm

Process: _____

Description: _____

_____

_____

_____

_____

_____

Rib cage

Lung

Diaphragm

Process: _____

Description: _____

_____

_____

_____

_____

_____

2. Draw an arrow indicating the direction of air flow through the trachea in each illustration.

### Graphing Data

A chemical analysis of the gases that are inhaled and exhaled is illustrated in the graph below.

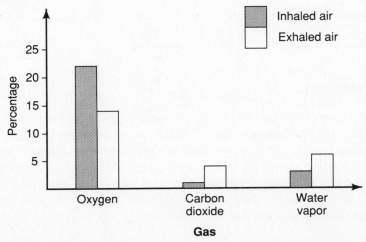

1. Based on the graph, describe what happens to the air that is inhaled.

_____

_____

_____

2. Where in the respiratory system does each of the events described in your answer to

question 1 occur? _____

_____

_____

_____

### Relating Structure and Function: Using the Main Ideas

1. How do the cilia and mucus found in the nose, pharynx, and trachea help these

structures perform their functions? _____

_____

_____

2. The trachea is made of strong rings of tough cartilage. What would happen if the trachea did

not contain cartilage? _____

_____

_____

3. Alveoli are made of elastic fibers that stretch during inhalation and pull back during
exhalation. How does the structure of the alveoli help the respiratory system carry

out gas exchange? _____

_____

_____

_____

### Concept Mapping

The construction of and theory behind concept mapping are discussed on
pages vii–ix in the front of this Study Guide. Read those pages carefully. Then
consider the concepts presented in Section 40–2 and how you would organize
them into a concept map. Now look at the concept map for Chapter 40 on
page 392. Notice that the concept map has been started for you. Add the key
facts and concepts you feel are important for Section 40–2. When you have
finished the chapter, you will have a completed concept map.

**Section
40-3**   **Control of the Respiratory System**   *(pages 890–891)*

## SECTION REVIEW

As a child, did you ever try to hold your breath until you got your way? In this section you learned why you could not do it for long! The respiratory system is under the direct control of the medulla oblongata in the lower part of the brain. Motor and sensory neurons are constantly monitoring the breathing muscles, as well as the amount of certain gases in the blood, to make sure that the cells of the body are getting the oxygen they need.

Special sensory receptors in important arteries monitor the acidity of the blood to determine the amount of oxygen getting to the cells. The acidity of the blood increases when there is too much carbon dioxide and not enough oxygen to allow cells to carry out their functions. When the acidity of the blood gets too high, the medulla oblongata takes over

and directs your breathing—even if you wanted to hold your breath, you couldn't! The control of the respiratory system by the medulla oblongata ensures that you·do not "forget" to breathe!

The lungs hold approximately 6 liters of air. Under normal conditions, you probably breathe 12 to 15 times a minute and exchange only about 0.6 liter of air with each breath. But the rate at which you breathe and the amount of air exchanged is related to the type of activities you are performing. During periods of heavy activity, the breathing rate and the amount of air exchanged with each breath increases to meet the body's demand for oxygen. The maximum amount of air that you can move into and out of your lungs is called vital capacity.

### Holding Your Breath: Using the Main Ideas

**1.** What part of the body controls the respiratory system? Why?

_____

_____

_____

**2.** What important blood factor determines how long you can hold your breath?

_____

_____

_____

**3.** Athletes who participate in synchronized swimming have to be able to hold their breath for long periods of time. Since the body must have a certain amount of oxygen in order to function, how have these athletes trained their bodies to meet their need for oxygen and yet be able to hold their breath for extended periods of time?

_____

_____

_____

_____

### Graphing Data

Use the graphs below to answer the following questions.

**1.** Which graph represents a person's breathing rate at rest? Explain your answer.

_____

_____

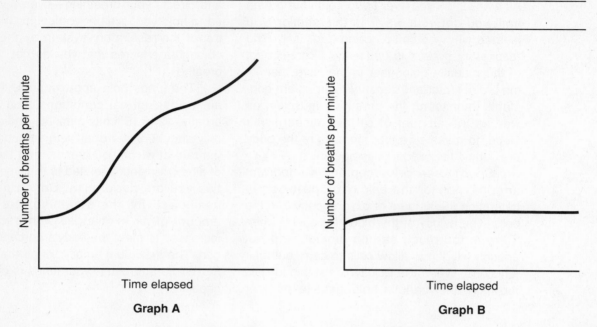

Graph A                     Graph B

**2.** On the graph below, draw a line that shows the relationship between the amount of exercise a person does and that person's vital capacity.

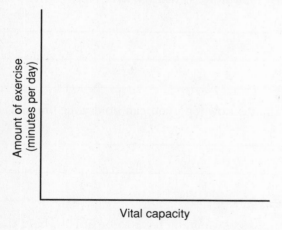

### Concept Mapping

The construction of and theory behind concept mapping are discussed on pages vii–ix in the front of this Study Guide. Read those pages carefully. Then consider the concepts presented in Section 40–3 and how you would organize them into a concept map. Now look at the concept map for Chapter 40 on page 392. Notice that the concept map has been started for you. Add the key facts and concepts you feel are important for Section 40–3. When you have finished the chapter, you will have a completed concept map.

## Using the Writing Process

Use your writing skills and imagination to respond to the following writing assignment. You will probably need an additional sheet of paper to complete your response.

One day a group of young red blood cells decided they did not like taking out the trash ($CO_2$) for all the other cells. They decided to form the "oxygen only" club, insisting that they would carry only oxygen. This caused a major upheaval throughout the body. You, as an aging red blood cell, have been chosen to confront these young red blood cells with the irresponsibility of their actions—and to persuade them to change their decision.

_____

_____

_____

_____

_____

_____

_____

_____

_____

_____

_____

_____

_____

_____

_____

_____

_____

_____

_____

_____

## Concept Mapping

The concept map below has been started for you. Add the key facts and concepts for each section of the chapter to this partial concept map. When you are done, you will have a concept map for the entire chapter.

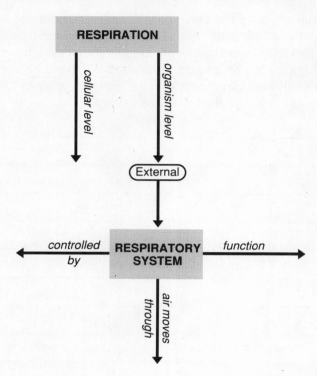

## S T U D Y
## G U I D E

CHAPTER
### Circulatory and Excretory Systems 41

### SECTION REVIEW

In this section you learned about the circulatory system. Composed of blood, blood vessels, and heart, the organs of the circulatory system work together to carry needed materials to and waste materials away from all the cells in the body. The heart can be described as two pumps —right and left—sitting side by side. Each of these pumps is divided into an upper chamber and a lower chamber. The upper chambers are called atria, and the lower chambers are called ventricles. Blood flows toward the heart in blood vessels called veins. A large vein filled with blood containing waste materials empties into the right atrium of the heart. When the pacemaker, found in the right atrium, begins the contraction of the heart muscle, the blood in the right atrium is squeezed into the right ventricle. Valves between each chamber prevent blood from going backward through the heart. When the heart contracts again, the blood in the right ventricle is squeezed into arteries that carry the blood away from the heart and to the lungs.

In the lungs, the waste gas carbon dioxide is exchanged for the oxygen required by the cells of the body. From the lungs, arteries carry this oxygen-rich blood back to the left atrium —the upper chamber on the left side of the heart. When the heart contracts, the blood is squeezed through another valve into the left ventricle. On the next heartbeat, blood from the left ventricle leaves the heart through the aorta, the main artery of the body, for its trip to deliver needed materials to all the cells of the body.

You also learned about the three types of blood vessels that serve as passageways for the blood in the circulatory system. Arteries carry blood away from the heart; veins carry blood toward the heart. As the aorta leaves the heart, it branches into smaller and smaller arteries that lead to the various organs of the body. Eventually, these small arteries branch into even smaller blood vessels called capillaries. Capillaries have such thin walls that the materials needed by the cells diffuse through their walls and across the cell membrane. At the same time, waste materials from the cells diffuse across the cell membrane and through the walls of the capillaries into the bloodstream. The blood containing waste materials is carried back toward the heart through larger and larger veins until it reaches the right atrium of the heart and once again begins its trip through the circulatory system.

### Circulatory System: Applying Vocabulary Skills

Identify the part or parts of the circulatory system described by each phrase.

**1.** Serve as passageways for the blood: _____

**2.** Made of cardiac muscle: _____

**3.** The upper chambers of the heart: _____

**4.** Located in the right atrium, this structure begins the heartbeat: _____

_____

5. Carry blood toward the heart: _____

6. Located between the upper and lower chambers of the heart, these structures prevent
   blood from flowing backward through the heart: _____

7. Lower chambers of the heart: _____

8. Thin-walled blood vessels through which materials diffuse into and out of individual
   cells: _____

9. Collects fluid lost by the blood in the body tissues and returns it to the circulatory
   system: _____

10. Type of blood vessels that carry blood away from the heart._____

### The Heart: Using the Main Ideas

1. Label the diagram of the heart shown below.

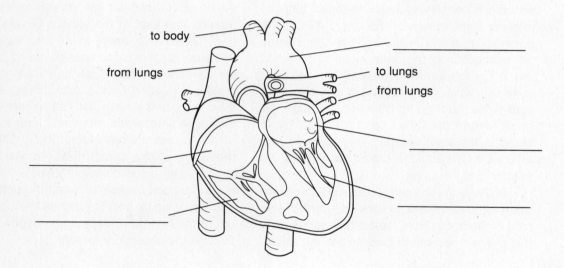

to body

from lungs

to lungs

from lungs

2. Using arrows, indicate the path that blood takes through the heart.

3. Complete the diagram by shading the side of the heart that contains deoxygenated blood.
   Use the following key to complete the diagram.

 Deoxygenated blood

 Oxygenated blood

### Blood Vessels: Interpreting Diagrams

Label the type of blood vessels shown in the diagram below.

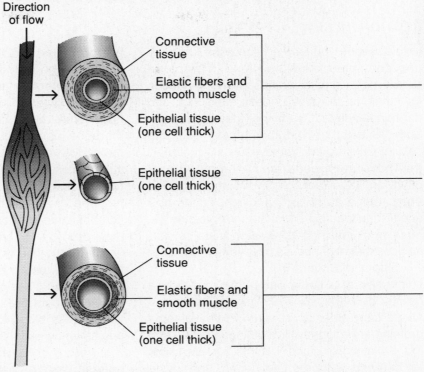

Direction of flow

Connective tissue

Elastic fibers and smooth muscle

Epithelial tissue (one cell thick)

Epithelial tissue (one cell thick)

Connective tissue

Elastic fibers and smooth muscle

Epithelial tissue (one cell thick)

### Identifying Pathways of Circulation

**1.** Which type of circulation carries blood to the lungs? _____

_____

**2.** Why is this type of circulation important? _____

_____

_____

_____

_____

**3.** Circulation that carries blood to all the major organs of the body is called

_____.

### Concept Mapping

The construction of and theory behind concept mapping are discussed on pages vii–ix in the front of this Study Guide. Read those pages carefully. Then consider the concepts presented in Section 41–1 and how you would organize them into a concept map. Now look at the concept map for Chapter 41 on page 402. Notice that the concept map has been started for you. Add the key facts and concepts you feel are important for Section 41–1. When you have finished the chapter, you will have a completed concept map.

| Section 41–2 | Blood | *(pages 905–907)* |

## SECTION REVIEW

In this section you learned about the functions and components of the blood. The blood serves as the major source of transportation for nutrients, dissolved gases, enzymes, and hormones needed by the individual cells of the body. As blood delivers these materials to the cells, it picks up and removes waste materials that have been generated. The blood also plays an important role in the regulation of body temperature, pH, and electrolyte balance. Blood is also important in protecting the body from invasion by foreign cells or substances that might cause infection.

The 4 to 6 liters of blood in the body is composed of a liquid part— plasma—and a cellular part. Plasma is 90 percent water and 10 percent plasma proteins. These plasma proteins are important in fighting infection and in the clotting of blood. Plasma is responsible for carrying nutrients, hormones, and waste products to and from the cells of the body.

The cellular part of the blood consists of three different types of cells. Red blood cells are produced in the bone marrow and are filled with the oxygen-carrying protein hemoglobin. Hemoglobin helps red blood cells carry enough oxygen to meet the demands of the body. White blood cells are also produced in the bone marrow, but they function to protect the body against infection from foreign cells or substances. When the body is invaded by a foreign substance, the white blood cells respond immediately to protect it. They can fight infection with chemicals they produce or they destroy the invader by phagocytosis ("eating" it!). The third cellular component of the blood is the platelets. When the skin is cut or scratched seriously enough to bleed, platelets begin the clotting process and prevent excess blood loss. The clotting process begins when platelets clump together to plug the wound or scratch. The platelets then release proteins called clotting factors that begin a series of chemical reactions designed to stop the bleeding.

### Functions of the Blood: Using the Main Ideas

Complete each sentence:

**1.** Blood transports _____

_____

_____

to and from the cells of the body.

**2.** Blood regulates _____

_____

_____ .

**3.** Blood protects the body from _____

_____

_____ .

### The Components of Blood: Applying Definitions

Identify the blood component that carries out each of the functions listed below.

**1.** Prevent blood loss by starting the clotting process: _____

_____

_____

**2.** Carry oxygen to the cells of the body: _____

_____

_____

**3.** Transports nutrients, hormones, enzymes, and waste materials to and from the cells of

the body: _____

_____

_____

**4.** Fight infection from foreign cells or substances: _____

_____

_____

### The Clotting Process: Sequencing Events

Place the events below in the proper sequence.

_____ Platelets release proteins called clotting factors.

_____ Platelets clump together to plug the wound.

_____ You cut your finger with a knife and begin to bleed.

_____ Clotting factors begin the chemical reactions that will complete the clotting process.

### Concept Mapping

The construction of and theory behind concept mapping are discussed on pages vii–ix in the front of this Study Guide. Read those pages carefully. Then consider the concepts presented in Section 41–2 and how you would organize them into a concept map. Now look at the concept map for Chapter 41 on page 402. Notice that the concept map has been started for you. Add the key facts and concepts you feel are important for Section 41–2. When you have finished the chapter, you will have a completed concept map.

**The Excretory System** *(pages 908–911)*

### SECTION REVIEW

In this section you learned that the excretory system removes waste materials from the body. The lungs eliminate the waste gas carbon dioxide, and the skin excretes excess water and salts in the form of sweat. But the main organs of the excretory system are the kidneys. The two kidneys function to filter the blood of excess water, urea (a toxic compound produced when amino acids are used for energy), and other waste materials. The kidneys keep the composition of the blood constant.

In this section you learned about the structure of the kidneys. Each kidney can be divided into two distinct regions: the renal medulla and the renal cortex. The renal cortex contains about 1 million nephrons. Nephrons are the structures that actually carry out the purification of the blood. As blood enters a nephron, it passes into a network of capillaries

called the glomerulus. The fluid from the blood diffuses through the capillary walls into Bowman's capsule. Impurities are filtered out and enter a collecting tubule. Much of the fluid that diffuses from the capillaries is reabsorbed by the blood. Reabsorption takes place in the loop of Henle, where important minerals such as sodium and potassium; nutrients such as amino acids, sugars, and fats; and water are removed from the fluid and reabsorbed through the walls of the capillaries and back into the bloodstream. The filtered waste material, urine, is concentrated in the loop of Henle, where a specialized system helps conserve water and minimize the amount of urine requiring excretion. Urine passes from the kidneys to the urinary bladder, where it is held until it is excreted by the body through the urethra.

### The Excretory System: Labeling Diagrams

**1.** Label the diagram of the excretory system shown below.

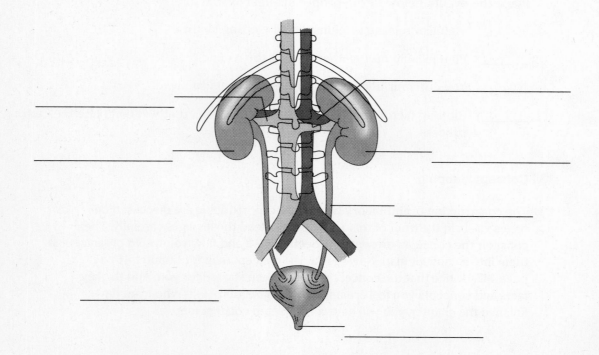

2. Use the diagram on the previous page to match the parts of the excretory system to their function.

**a.** Carries urine from the kidneys to the urinary bladder: _____

_____

**b.** Carries waste-laden blood into the kidneys: _____

_____

**c.** Purifies blood by removing excess water, urea, and other waste products:

_____

_____

**d.** Collects and stores urine until it is excreted from the body: _____

_____

**e.** Carries purified blood from the kidneys toward the heart: _____

_____

**f.** Provides a passageway for urine as it leaves the body: _____

_____

### The Kidneys: Relating Structures and Functions

1. Label the diagram of the kidney shown below.

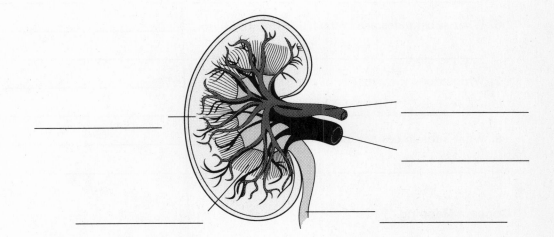

2. Which structure in the kidney actually carries out the purification process?

_____

**3.** Use the following terms to complete the accompanying diagram: *Filtration, Reabsorption, Arteriole, Venule, Bowman's capsule, Glomerulus, Loop of Henle,* and *Collecting tubule.*

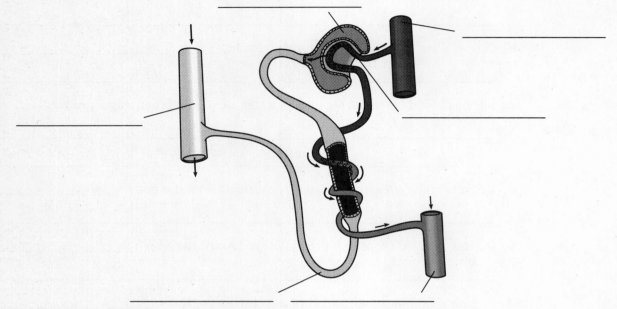

Use the diagram you have just completed to help answer the following questions.

**4.** Where does the process of filtration take place? _____

_____

**5.** What substances diffuse into Bowman's capsule? _____

_____

**6.** What substances are reabsorbed into the blood? _____

_____

**7.** Where does the process of reabsorption take place? _____

_____

**8.** What substances are excreted as urine? _____

_____

### Concept Mapping

The construction of and theory behind concept mapping are discussed on pages vii–ix in the front of this Study Guide. Read those pages carefully. Then consider the concepts presented in Section 41–3 and how you would organize them into a concept map. Now look at the concept map for Chapter 41 on page 402. Notice that the concept map has been started for you. Add the key facts and concepts you feel are important for Section 41–3. When you have finished the chapter, you will have a completed concept map.

## Using the Writing Process                                    *Chapter 41*

Use your writing skills and imagination to respond to the following writing
assignment. You will probably need an additional sheet of paper to
complete your response.

The circulatory system generally gets more attention than the lymphatic
system. Some people don't even know that the lymphatic system exists. Write
five suggestions as to how you could increase public awareness of the
importance of the lymphatic system. Implement at least one of your
suggestions.

_____

_____

_____

_____

_____

_____

_____

_____

_____

_____

_____

_____

_____

_____

_____

_____

_____

_____

_____

_____

_____

**Concept Mapping** | *Chapter 41*

The concept map below has been started for you. Add the key facts and concepts for each section of the chapter to this partial concept map. When you are done, you will have a concept map for the entire chapter.

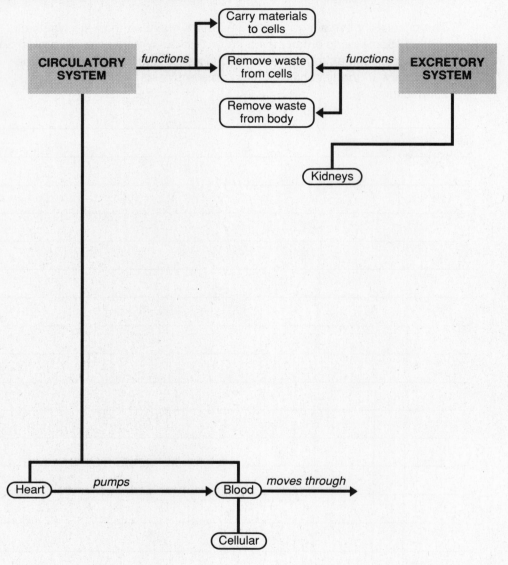

|

S T U D Y
G U I D E

CHAPTER
*Endocrine System* **42**

**Endocrine Glands** *(pages 917–926)*

### SECTION REVIEW

In this section you learned about the endocrine system and the glands that make up this system. The endocrine system regulates many of the body's activities by secreting hormones directly into the bloodstream.

Hormones are chemical messengers that travel in the bloodstream from the endocrine gland that produced them to target cells in other parts of the body. The hormones attach to the target cells and affect their functions.

In this section you also learned about the major endocrine glands, where they are located, the hormones they produce, and the effect these hormones have on the body. The thyroid gland, located in the throat, produces the hormone thyroxine, which regulates the metabolic rate, the rate at which the cells produce energy, in the cells. The thyroid also releases calcitonin, which regulates the amount of calcium in the blood. Located within the thyroid are four small glands called the parathyroids. The parathyroid glands produce and release parathyroid hormone (PTH), which also helps regulate the amount of calcium in the blood.

The adrenal glands located on top of the kidneys are divided into two parts: the adrenal cortex and the adrenal medulla. The adrenal cortex, the outer part of each adrenal gland, produces and secretes aldosterone, which affects the water and salt balance in the body by stimulating the kidneys to reabsorb sodium and secrete potassium. The corticosteroids produced by the adrenal cortex help control the metabolism of carbohydrates, fats, and proteins. The adrenal medulla secretes two hormones that prepare the body for "fight or flight." When you are faced with a stressful situation, the adrenal medulla releases adrenaline and noradrenaline (sometimes called epinephrine). Adrenaline increases the heart rate, blood pressure, metabolism, and the amount of blood flowing to the skeletal muscles (in case you have to run!). At the same time, noradrenaline stimulates the heart muscle. These two hormones increase the activity going on throughout the body so that you are better able to respond to the situation.

The reproductive glands are the ovaries in females and the testes in males. In addition to producing the sex cells, eggs and sperm, the gonads produce hormones that support the development of the sex cells and the physical characteristics associated with each sex.

The islets of Langerhans, located in the pancreas, produce and secrete insulin and glucagon. These two hormones work together to regulate the amount of sugar in the bloodstream.

In this section you also learned about the pituitary gland and the hypothalamus. The pituitary gland is located at the base of the skull and is divided into two sections: the anterior pituitary and the posterior pituitary. Some of the hormones produced by the anterior pituitary control the production and release of hormones in other endocrine glands: thyroid-stimulating hormone (TSH) controls the thyroid gland, adrenocorticotropic hormone (ACTH) controls the adrenal cortex, and follicle-stimulating hormone (FSH) and luteinizing hormone (LH) control the reproductive glands. Several other hormones secreted by the anterior pituitary are the growth hormone (GH), which controls normal growth in the cells of the body; prolactin, which stimulates milk production in pregnant women; and melanocyte-stimulating hormone (MSH), which stimulates the production of pigment in the melanocytes of the skin. The posterior pituitary produces antidiuretic hormone (ADH), which stimulates the kidneys to reabsorb water from urine, and oxytocin, which stimulates the muscles in the uterus during childbirth and the release of milk from the breasts of nursing mothers. The hypothalamus is attached to the posterior pituitary gland. Based on sensory information from the nervous system, the hypothalamus controls the secretions of the pituitary gland.

▨ **The Endocrine System: Applying Definitions**

1.  What is the function of the endocrine system? _____

    _____

    _____

    _____

2.  What are hormones?_____

    _____

    _____

3.  How do hormones carry out their functions? _____

    _____

    _____

    _____

▨ **The Endocrine System: Interpreting Diagrams**

Identify and label the endocrine glands shown in the diagram below.

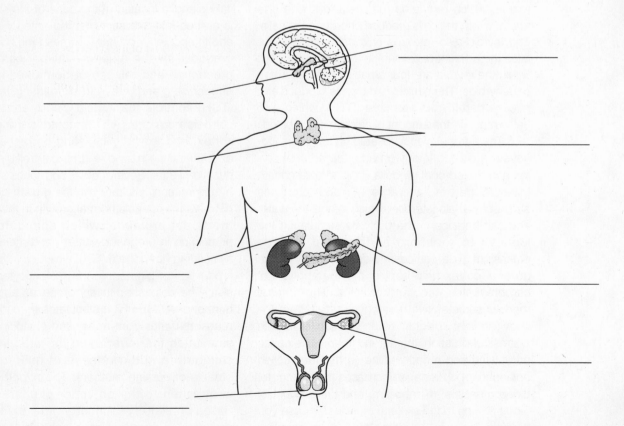

### Functions of the Endocrine Glands: Using the Main Ideas

Complete the chart.

| Endocrine Gland | Hormone | Effect on Target Cells |
|---|---|---|
| | Thyroxine | |
| | | Regulates calcium levels in the blood |
| Parathyroid | | |
| | | Affect water and salt balance and metabolism of carbohydrates, fats, and proteins |
| | Adrenaline and noradrenaline | |
| Ovaries | Estrogen | |
| | | Prepares the uterus for the developing embryo |
| | Androgens | |
| | | Regulate the amount of sugar in the bloodstream |
| Hypothalamus | Releasing hormones | |
| | | Stimulates kidneys to reabsorb water |
| | | Stimulates contractions of uterus during childbirth |
| Anterior pituitary | | Stimulate other endocrine glands to produce and secrete their hormone |
| | | Stimulates milk production in pregnant women |
| | Growth hormone | |
| | Melanocyte-stimulating hormone | |

### Concept Mapping

The construction of and theory behind concept mapping are discussed on pages vii–ix in the front of this Study Guide. Read those pages carefully. Then consider the concepts presented in Section 42–1 and how you would organize them into a concept map. Now look at the concept map for Chapter 42 on page 410. Notice that the concept map has been started for you. Add the key facts and concepts you feel are important for Section 42–1. When you have finished the chapter, you will have a completed concept map.

## Section 42–2   Control of the Endocrine System        *(pages 927–929)*

### SECTION REVIEW

In the last section you learned that the endocrine system controls many functions in the body. In this section you learned that, like other body systems, the endocrine system is regulated by a negative-feedback mechanism. Negative feedback means that a stimulus (in this case, the amount of hormone in the blood) produces a reaction that ultimately reduces the stimulus. Within the endocrine system, the hormone output of the different endocrine glands is constantly monitored by the hypothalamus. The amount of the hormone in the bloodstream determines whether or not the gland should continue to secrete the hormone. If there is too little of a particular hormone in the bloodstream, the hypothalamus sends releasing hormones to the anterior pituitary. This action causes the anterior pituitary to release a hormone that will stimulate the correct endocrine gland to produce the needed hormone. If there is too much of a particular hormone in the bloodstream, the opposite occurs.

You also learned that the hormones produced by the endocrine system can be divided into two groups based on the way in which they carry out their action. Polypeptide hormones bind to receptors on the cell membrane of their target cells. This action activates enzymes that act as messengers to carry out the action of the hormone and to alter the activity of the cell. Steroid hormones diffuse through the cell membrane and attach to receptor molecules. With their receptor molecules, steroid hormones travel to the cell's nucleus and attach to certain gene sequences to control gene expression directly.

In this section you also learned about "local" hormones called prostaglandins. Prostaglandins are called local hormones because they function only within the cells in which they are produced. Prostaglandins are involved in the contractions of smooth muscles and in the sensation of pain. Prostaglandins have helped scientists explain how aspirin works: It stops the production of pain-causing prostaglandins.

### Control of the Endocrine System: Using the Main Ideas

1. Why is the system that controls the endocrine system called a negative-feedback system?

_____

_____

_____

_____

2. Which endocrine gland receives information about the levels of hormones in the bloodstream? _____

_____

_____

### Negative-Feedback Mechanism: Interpreting Diagrams

Number the arrows in the diagram to correctly match the sequence of events in this negative-feedback system.

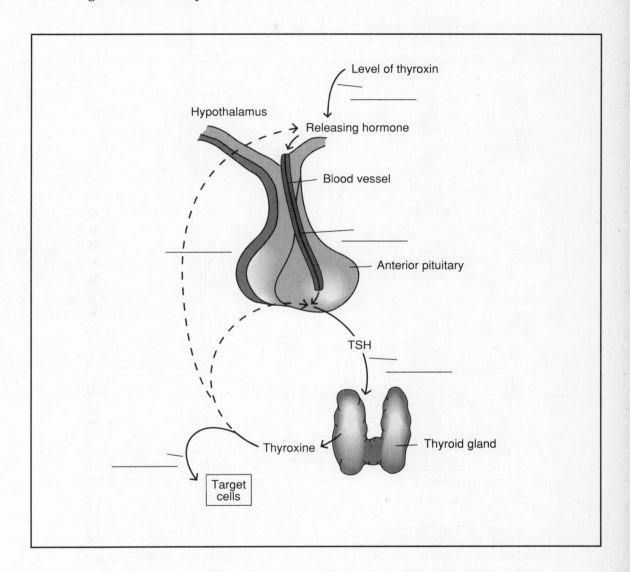

1. The hypothalamus is activated by a low level of thyroxin in the bloodstream.

2. The hypothalamus produces a releasing hormone that acts on the anterior pituitary.

3. The anterior pituitary releases thyroid-stimulating hormone into the bloodstream.

4. Thyroid-stimulating hormone stimulates the thyroid to release thyroxin into the bloodstream, which increases the rate of metabolism in the cells of the body.

5. The increased level of thyroxin in the bloodstream causes the hypothalamus to stop producing the releasing hormone.

### Hormone Action: Applying Definitions

1. The diagrams below represent two ways in which hormones carry out their actions. In the spaces provided, identify the type of hormone illustrated in each diagram.

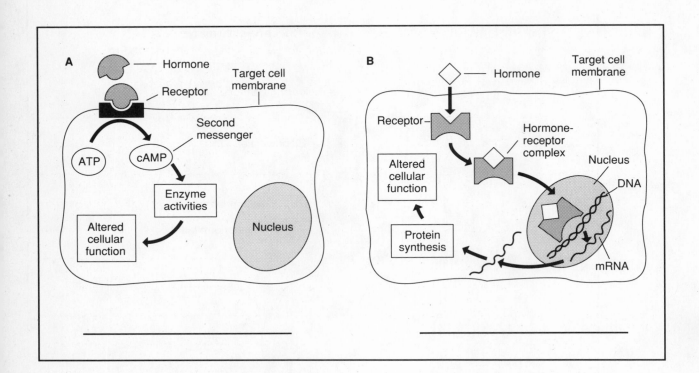

2. _____ are hormones that carry out their actions in the cells in which they are produced.

### Concept Mapping

The construction of and theory behind concept mapping are discussed on pages vii–ix in the front of this Study Guide. Read those pages carefully. Then consider the concepts presented in Section 42–2 and how you would organize them into a concept map. Now look at the concept map for Chapter 42 on page 410. Notice that the concept map has been started for you. Add the key facts and concepts you feel are important for Section 42–2. When you have finished the chapter, you will have a completed concept map.

## Using the Writing Process

Use your writing skills and imagination to respond to the following writing assignment. You will probably need an additional sheet of paper to complete your response.

The pituitary gland wants to specialize by producing only three types of hormones. The gland has come to you for advice as to which three hormones it should produce. Discuss the advice you would provide for the gland, keeping in mind the needs of both the pituitary and the society (the body) it serves.

_____

_____

_____

_____

_____

_____

_____

_____

_____

_____

_____

_____

_____

_____

_____

_____

_____

_____

_____

_____

_____

_____

**Concept Map**                                                    *Chapter 42*

The concept map below has been started for you. Add the key facts and concepts for each section of the chapter to this partial concept map. When you are done, you will have a concept map for the entire chapter.

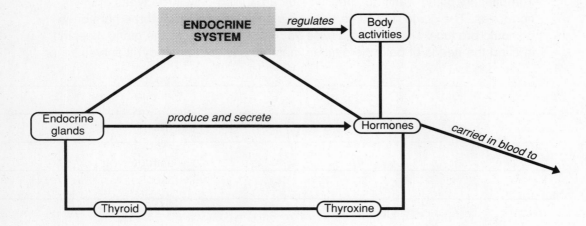

S T U D Y
G U I D E

**Section 43–1**  **The Reproductive System**  *(pages 935–943)*

## SECTION REVIEW

In this section you learned about the reproductive system, which functions to produce, store, nourish, and release sex cells called gametes. When gametes are fertilized, new individuals result and the species survives.

You also learned about sexual development in the human embryo. Male and female embryos are identical for the first 6 weeks of development. But during the seventh week after fertilization, the main organs of the reproductive system—testes in males and ovaries in females—begin to produce hormones that determine whether the embryo will develop into a male or a female. The ovaries and the testes continue to produce small amounts of hormones through birth and until the individual reaches puberty—between the ages of 9 and 15. During puberty, the reproductive system becomes functional and sexual maturity is reached.

In this section you also learned that within the seminiferous tubules of the testes, sperm, the male gametes, and the male hormone testosterone are produced. Sperm mature and are stored in the epididymis, from which they pass into the vas deferens in the abdomen. In the vas deferens, seminal fluid is added to the sperm, producing semen. From the vas deferens, sperm pass through the urethra to the outside of the body through the penis.

Within the ovaries, which are the main organs of the female reproductive system, an ovum (or egg)—the female gamete—is produced and released each month. The ovaries also produce estrogens, the hormones responsible for the secondary characteristics of the female. The negative-feedback mechanism of the endocrine system and the reproductive system works to produce a series of events called the menstrual cycle. Controlled by hormones, the menstrual cycle involves the development and release of an egg by an ovary and the preparation of the uterus to receive the egg if it has been fertilized. The menstrual cycle is divided into four phases: the follicle phase, ovulation, the luteal phase, and menstruation.

You also learned that in order for fertilization to occur, sperm must be present in a Fallopian tube at the same time as an egg. Once an egg is fertilized, its cell membrane changes to prevent any other sperm cells from entering. The fertilized egg is now a zygote.

### The Reproductive System: Using the Main Ideas

**1.** What is the function of the reproductive system? _____

_____

_____

**2.** Is the reproductive system necessary for the survival of an individual? For the survival

of a species? Explain your answer. _____

_____

_____

### Sexual Development: Sequencing Events

Use the numbers 1 through 4 to correctly sequence the following phrases, which describe human sexual development.

_____ During the seventh week, the sex organs produce hormones that direct the development of the male or the female reproductive system.

_____ Human embryos are identical.

_____ During puberty, reproductive systems become functional and sexual maturity is reached.

_____ Reproductive organs continue to produce small amounts of male and female hormones.

### The Male Reproductive System: Identifying Structures and Functions

**1.** Complete the diagram by labeling the organs of the male reproductive system.

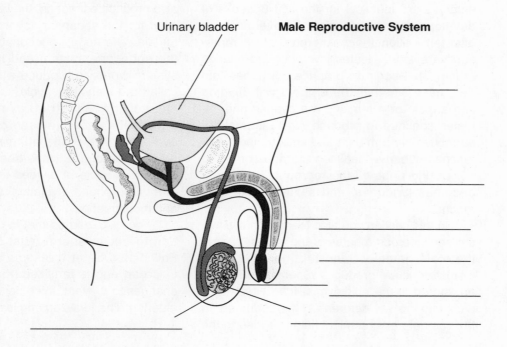

Urinary bladder     **Male Reproductive System**

**2.** Identify the organ of the male reproductive system that performs each of the following functions.

    **a.** Sac that holds the testes outside the body: _____

    **b.** Produce sperm cells: _____

    **c.** Produce testosterone: _____

    **d.** Stores sperm until they are mature: _____

    **e.** Passageway through the penis: _____

    **f.** Passes through the abdomen where seminal fluid is added to the sperm to produce semen: _____

**3.** In the space provided, draw and label a male sex cell.

### ▨ The Female Reproductive System: Identifying Structures and Functions

**1.** Complete the diagram by labeling the organs of the female reproductive system.

**Female Reproductive System**

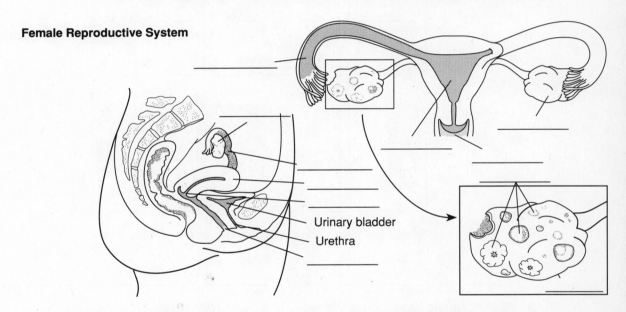

Urinary bladder

Urethra

**2.** Identify the organ of the female reproductive system that performs each of the following functions.

   **a.** Produces estrogen: _____

   **b.** Prepare an ovum for ovulation: _____

   **c.** Prepares itself to nourish and protect a fertilized egg: _____

   **d.** Passageway through which the egg travels on its trip to the uterus: _____

   **e.** Site of fertilization: _____

   **f.** Canal leading outside the body from the uterus: _____

**3.** What are the two functions of the female reproductive system?_____

_____

**4.** How does the endocrine system control the activities of the female reproductive system?

_____

_____

### The Menstrual Cycle: Interpreting Diagrams

**1.** Use the terms listed below to correctly complete the diagram. Then use the correctly labeled diagram to answer the questions that follow.

ovulation       menstruation       luteal phase       follicle phase

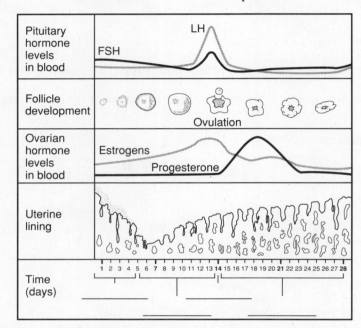

**2.** What happens to the levels of FSH, LH, and estrogens in the blood during ovulation?

_____

**3.** What hormone level increases following ovulation? _____

**4.** During menstruation, what happens to the uterine lining? _____

_____

**5.** In terms of the follicle, when does ovulation occur? _____

### Concept Mapping

The construction of and theory behind concept mapping are discussed on pages vii–ix in the front of this Study Guide. Read those pages carefully. Then consider the concepts presented in Section 43–1 and how you would organize them into a concept map. Now look at the concept map for Chapter 43 on page 418. Notice that the concept map has been started for you. Add the key facts and concepts you feel are important for Section 43–1. When you have finished the chapter, you will have a completed concept map.

| Section 43–2 | **Human Development** | *(pages 943–947)* |

## SECTION REVIEW

In this section you learned how a fertilized egg, or zygote, develops from a single cell into a multicellular human being. After fertilization in a Fallopian tube, the zygote immediately begins to grow through a series of cell divisions. About 4 days after fertilization, the zygote consists of about 50 cells and is called a morula. As cell division continues, a cavity filled with fluid develops and becomes a hollow structure called a blastocyst. About 6 or 7 days after fertilization, the blastocyst implants itself in the nourishing lining of the uterus. Membranes called the amnion and the chorion also form from the blastocyst in order to protect the embryo. Within the cavity of the blastocyst, clusters of cells divide into three layers—the ectoderm, mesoderm, and endoderm—in the process of gastrulation. From these three layers will develop all the organs and tissues of the embryo.

At the end of the third week of development, the nervous and digestive systems have begun to form, and the chorion membrane has grown into the placenta. The placenta serves as a connection between mother and developing embryo: a supply system for nutrients needed by the developing embryo and an excretory system for the wastes produced from its cells. After 8 weeks of development, the embryo is called a fetus. For the next 7 months, the fetus will be protected and nourished in the mother's uterus while its organ systems continue to develop so that they can support independent life.

In this section you also learned that at the end of 9 months, when the fetus is ready for life on its own, childbirth occurs. The endocrine system begins the process of childbirth when the pituitary gland releases the hormone oxytocin. Oxytocin causes the smooth muscle cells of the uterus to begin a series of contractions called labor. These contractions enlarge the cervix, or the uterus's opening into the vagina, so that the baby can pass out of the uterus, into the vagina, and out into the world.

### Stages of Development: Sequencing Events

1. Use the numbers 1 through 8 to correctly sequence the stages of development illustrated below.

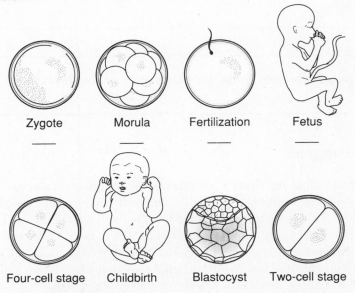

Zygote     Morula     Fertilization     Fetus

_____     _____     _____     _____

Four-cell stage     Childbirth     Blastocyst     Two-cell stage

_____     _____     _____     _____

Now use the correctly completed figure to answer the following questions.

**2.** Which stages occur in the Fallopian tubes? _____

**3.** Which structure implants itself in the uterus? _____

**4.** At which stage does gastrulation occur? _____

**5.** Describe gastrulation and its importance. _____

_____

_____

**6.** After 3 weeks of development, the _____, which serves as a connection between mother and developing embryo, forms. The substances that pass

between mother and embryo are _____

_____ .

**7.** After 8 weeks of development, the embryo is called a _____ .

**8.** After _____ months of development, childbirth occurs.

**Childbirth: Using the Main Ideas**

**1.** What two body systems work together in the process of childbirth?

_____

**2.** Which system begins the process of childbirth? _____

**3.** How does this system begin the process of childbirth? _____

_____

**4.** What is labor? _____

_____

_____

**5.** After a baby is born, what structures are expelled from the uterus?

_____

**Concept Mapping**

The construction of and theory behind concept mapping are discussed on pages vii–ix in the front of this Study Guide. Read those pages carefully. Then consider the concepts presented in Section 43–2 and how you would organize them into a concept map. Now look at the concept map for Chapter 43 on page 418. Notice that the concept map has been started for you. Add the key facts and concepts you feel are important for Section 43–2. When you have finished the chapter, you will have a completed concept map.

## Using the Writing Process

Use your writing skills and imagination to respond to the following writing assignment. You will probably need an additional sheet of paper to complete your response.

Design a poster or advertising campaign whose main theme is, "There's a lot more to having a baby than just knowing how."

_____

_____

_____

_____

_____

_____

_____

_____

_____

_____

_____

_____

_____

_____

_____

_____

_____

_____

_____

_____

## Concept Mapping

The concept map below has been started for you. Add the key facts and concepts for each section of the chapter to this partial concept map. When you are done, you will have a concept map for the entire chapter.

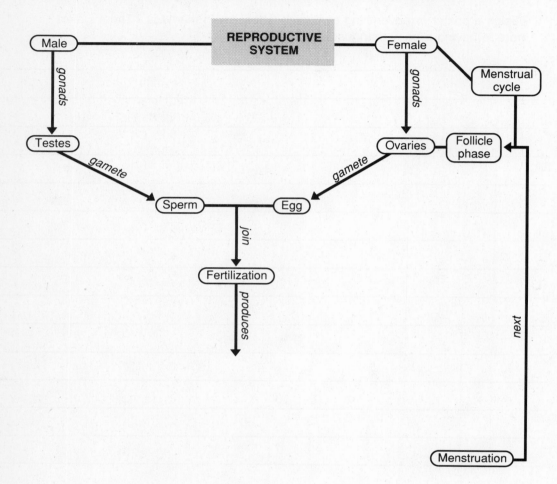

© Prentice-Hall, Inc.

S T U D Y
G U I D E

CHAPTER
*Human Diseases* **44**

<table>
<tr><td>**Section**<br>**44–1**</td><td>**The Nature of Disease**</td><td>*(pages 953–955)*</td></tr>
</table>

### SECTION REVIEW

In this section you learned that any change other than an injury that interferes with your body's normal functioning is called a disease. Infectious diseases are caused when pathogens such as bacteria, viruses, or other disease-causing microorganisms invade the body.

Infectious diseases can be spread in a variety of ways. Some pathogens cause infection by entering the body through breaks in the skin. Some diseases are spread from one person to another through coughing, sneezing, or sexual contact. Sometimes diseases are spread when the water or food supplies become contaminated with disease-causing microorganisms or when a person is bitten by an animal that has been infected.

In this section you also learned about the work of Louis Pasteur and Robert Koch. In the nineteenth century, these men developed the germ theory of infectious disease. Their theory stated that diseases are caused by micro-

organisms—not evil spirits, magic, or bad luck! Koch, as he continued his study of infectious diseases, developed a set of rules for proving which of the millions of microorganisms in a host organism was the cause of a specific disease. These rules are called Koch's postulates, and they state that

- The microorganism is present in the diseased or host organism but not in a healthy organism.
- The microorganism must be isolated and grown away from the host.
- When injected into a new host organism, the culture-grown microorganism must produce the disease.
- The same microorganism should be reisolated from the second host and grown in a culture. The microorganism should be the same as the original microorganism.

Robert Koch's postulates are used by doctors today to determine the cause of infectious diseases.

### Infectious Diseases: Applying Vocabulary Skills

Read the following situation carefully and then answer the questions that follow.

Pam's little sister had been home sick all week with a sore throat and a fever. On Friday, Pam had a date with Henry to go see a movie. At the movie Pam and Henry shared a soda and a box of popcorn. On Saturday, Pam had no energy and spent most of the day watching TV and napping. By Sunday, she had developed a terrible sore throat and was running a fever of 101°.

**1.** Did Pam have a disease? How do you know? _____

_____

**2.** What symptoms did Pam exhibit? _____

**3.** How did Pam get infected with the microorganism that caused her symptoms?

_____

4. The disease-causing microorganism that invaded Pam's body is called a _____.

5. Henry was exposed to the disease because of his _____ with Pam.

   But Henry will get sick only if his body has been _____ with the

   pathogen and it begins to injure his cells and tissues.

6. Is Pam's disease infectious or noninfectious? _____

### The Germ Theory of Disease: Using the Main Ideas

1. Who developed the germ theory of disease? _____

2. What does the germ theory of disease say about the cause of disease?

   _____

3. What do Koch's postulates allow scientists to determine? _____

   _____

4. Use Koch's postulates to answer the following questions:

   a. Where will a disease-causing organism always be found? _____

      _____

   b. Why is the microorganism grown in a pure culture away from the host?

      _____

      _____

   c. What happens when the cultured microorganism is injected into a new host?

      _____

   d. Why is the microorganism isolated and cultured from the second host?

      _____

5. What conclusions might a scientist draw if the isolated and cultured microorganism does

   not produce the disease in the new host organism?_____

   _____

### Concept Mapping

The construction of and theory behind concept mapping are discussed on
pages vii–ix in the front of this Study Guide. Read those pages carefully. Then
consider the concepts presented in Section 44–1 and how you would organize
them into a concept map. Now look at the concept map for Chapter 44 on
page 426. Notice that the concept map has been started for you. Add the key
facts and concepts you feel are important for Section 44–1. When you have
finished the chapter, you will have a completed concept map.

## SECTION REVIEW

In this section you learned that infectious diseases are divided into groups based on the kind of pathogen that causes them. Viruses and bacteria are the most common pathogens, but rickettsiae, fungi, and protozoans also cause infectious diseases.

Viruses are noncellular particles that invade living cells. Once inside the host cell, the viral DNA or RNA takes over the functions of the cell and produces new viral particles that can infect other cells and destroy the infected cell. Many viral diseases are spread through direct contact with an infected person. Examples of viral diseases are AIDS, polio, smallpox, measles, and the common cold.

You also learned that the bacteria that cause disease do so in several ways. The bacteria may infect the tissues directly or they may produce toxins, or poisons, that cause illness. Bacterial diseases can be spread by person-to-person contact, exposure to the bacterium, or by contaminated food products. Tuberculosis (TB), pneumonia, and tetanus are examples of bacterial diseases.

Rickettsiae are disease-causing pathogens that, like viruses, grow only within living cells. Rickettsiae are transmitted by the bite of infected ticks, lice, or fleas. Examples of diseases caused by rickettsiae are Rocky Mountain spotted fever and typhus.

You also learned that a few fungi can cause diseases. These diseases are caused by fungi normally found on the skin. But when the fungi grow rapidly, they can cause serious infections. The most common fungal diseases are athlete's foot and ringworm.

This section also introduced you to the thirty or so protozoans that can cause human diseases. Protozoan infections are most common in the tropical regions of the world, where warm, moist climates provide protozoans with the conditions required for their survival.

### Diseases You Have Had: Applying the Main Ideas

Using three different infectious diseases you have had, complete the chart below.

| Disease | Agent of Disease | Symptoms |
|---------|------------------|----------|
|         |                  |          |
|         |                  |          |
|         |                  |          |

■ **Classifying Infectious Diseases: Using the Main Ideas**

1. How are infectious diseases classified? _____

_____

_____

2. Based on your answer to question 1, divide the following diseases into groups.

   Measles, ringworm, Lyme disease, TB, malaria, chicken pox, pneumonia,
   African sleeping sickness, athlete's foot, typhus, Rocky Mountain spotted
   fever, amebic dysentery, diphtheria

_____

_____

_____

_____

_____

_____

_____

_____

_____

_____

_____

■ **Concept Mapping**

The construction of and theory behind concept mapping are discussed on
pages vii–ix in the front of this Study Guide. Read those pages carefully. Then
consider the concepts presented in Section 44–2 and how you would organize
them into a concept map. Now look at the concept map for Chapter 44 on
page 426. Notice that the concept map has been started for you. Add the key
facts and concepts you feel are important for Section 44–2. When you have
finished the chapter, you will have a completed concept map.

**Section 44-3**  **Cancer**  *(pages 960-963)*

## SECTION REVIEW

In this section you learned about the noninfectious disease called cancer. Cancer is a life-threatening disease caused when the body's own cells multiply uncontrollably and destroy healthy tissue. When the body's cells begin to grow and divide out of control, a mass of tissue called a tumor forms. Tumors can be benign or malignant. Malignant tumors may spread to other parts of the body and absorb the nutrients needed by cells to function properly.

In this section you also learned that scientists do not know all the causes of cancer. They do know that repeated and prolonged exposure to cancer-causing substances can cause a cancer to develop. According to the viral theory of cancer, cancer-causing genes in viruses, called oncogenes, use their genetic material to upset the normal balance of cell growth.

Radiation can also cause cancer because it produces mutations, or changes, in the DNA structure of cells. If these mutations affect the genes that control cell growth, a cancer cell may result. Radiation that can cause cancer includes nuclear radiation, X-rays, and ultraviolet radiation. Overexposure to the ultraviolet radiation in sunlight is the main cause of skin cancer. The third main cause of cancer is cancer-causing chemicals called carcinogens. Like radiation, carcinogens cause cancer by producing mutations in normal cells. Carcinogens can be natural or synthetic compounds.

You also learned that once cancer has been detected, doctors have three main methods of treatment: surgery, radiation therapy, and drug therapy. Surgery removes a cancerous tumor from the body so that it can do no more damage. If a cancerous tumor has been detected early and can be surgically removed, the patient has a good chance of complete recovery. Radiation therapy is used to destroy the faster growing cancer cells before they can spread to other parts of the body. Drug therapy, called chemotherapy, uses chemicals to destroy cancer cells. Often these three types of treatment are used together to destroy harmful cancer cells. But the best weapon against cancer is early detection!

### Cancer: Using the Main Ideas

**1.** What is cancer? _____

_____

_____

_____

_____

**2.** Why is cancer considered a disease? _____

_____

_____

_____

_____

_____

**3.** Is cancer an infectious or a noninfectious disease? Explain your answer.

_____

_____

_____

_____

_____

_____

**4.** What advice would you give a person who was trying to lower the chances of getting

cancer?_____

_____

_____

_____

_____

_____

_____

**5.** If cancer is detected in a person, what types of treatment might a doctor use?

_____

_____

_____

_____

_____

### Concept Mapping

The construction of and theory behind concept mapping are discussed on pages vii–ix in the front of this Study Guide. Read those pages carefully. Then consider the concepts presented in Section 44–3 and how you would organize them into a concept map. Now look at the concept map for Chapter 44 on page 426. Notice that the concept map has been started for you. Add the key facts and concepts you feel are important for Section 44–3. When you have finished the chapter, you will have a completed concept map.

**Using the Writing Process**                                    *Chapter 44*

Use your writing skills and imagination to respond to the following writing assignment. You will probably need an additional sheet of paper to complete your response.

Write a short story or play or skit about a future time in which all disease-causing organisms, including bacteria and fungi, have been wiped out. Keep this in mind, however: Such a future might not be as healthful as it sounds.

_____

_____

_____

_____

_____

_____

_____

_____

_____

_____

_____

_____

_____

_____

_____

_____

_____

_____

_____

_____

_____

## Concept Mapping

The concept map below has been started for you. Add the key facts and concepts for each section of the chapter to this partial concept map. When you are done, you will have a concept map for the entire chapter.

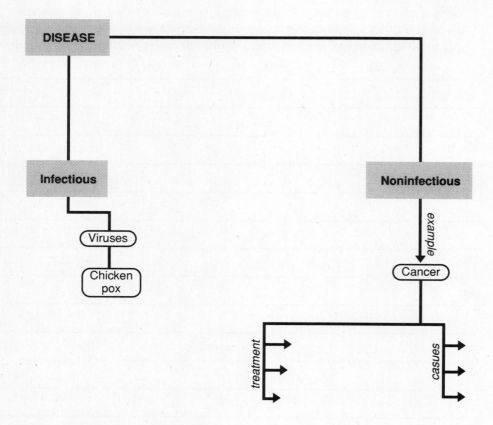

STUDY
GUIDE

| Section 45–1 | **Nonspecific Defenses** | *(pages 969–971)* |

### SECTION REVIEW

In this section you learned that the immune system provides the body with its primary defense against disease-causing microorganisms. The immune system has two types of defenses: nonspecific and specific. As their names indicate, nonspecific defenses defend the body against any disease-causing organism or pathogen; specific defenses work against a particular pathogen that causes a specific disease.

The most important of the body's three main nonspecific defenses consists of the skin and other barriers. They serve as the first line of defense against pathogens. The skin prevents pathogens from entering the body. The acidic environment of the skin kills many pathogens; hair and mucus secretions trap particles that try to invade the body through the nose and throat; and many of the body's secretions contain lysozyme, an enzyme that breaks down cell walls of bacteria.

The second line of nonspecific defenses, referred to as the inflammatory response, is necessary when large numbers of pathogens are successful in invading the body. Bacteria within a wound cause fluid and white blood cells to enter the area. Then phagocytes, a type of white blood cell, destroy the bacteria. The area may become inflamed, or red and swollen, as the phagocytes do their job. If the infection is serious enough and the pathogens have spread throughout the body, the immune system produces more white blood cells and releases chemicals that stimulate the white blood cells by increasing the body's temperature. This action produces a fever. Fever is a sure sign that the body is fighting an infection. Many pathogens can survive only in a very narrow temperature range. When you "have a fever," your higher body temperature is slowing down or stopping the growth of pathogens.

You also learned about interferon—the third nonspecific defense the body uses to fight disease. Interferons are produced by virus-infected cells and help other cells resist the viral infection. Interferons "interfere" with the process of viral replication and allow the specific defenses of the body time to respond to the infection.

### The Immune System: Using the Main Ideas

1. What is the function of the immune system? _____

   _____

   _____

2. The immune system = _____ + _____.

3. What is the difference between the types of defenses that make up the immune system?

   _____

   _____

### Nonspecific Defenses: Applying the Main Ideas

Identify the nonspecific defense(s) or structures described in each situation.

**1.** Trap bacteria, dust, and other foreign particles in the throat:

_____

_____

**2.** Produce an acidic environment on the surface of the skin that kills many pathogens:

_____

_____

**3.** Type of white blood cell that engulfs and destroys bacteria:

_____

_____

**4.** Most important nonspecific defense: _____

_____

**5.** Enzyme in body secretions that breaks down the cell wall of many bacteria:

_____

_____

**6.** Higher body temperature that slows down or stops the growth of pathogens:

_____

_____

### Concept Mapping

The construction of and theory behind concept mapping are discussed on pages vii–ix in the front of this Study Guide. Read those pages carefully. Then consider the concepts presented in Section 45–1 and how you would organize them into a concept map. Now look at the concept map for Chapter 45 on page 434. Notice that the concept map has been started for you. Add the key facts and concepts you feel are important for Section 45–1. When you have finished the chapter, you will have a completed concept map.

## SECTION REVIEW

In this section you learned about the specific defenses of the immune system. These defenses fight disease-causing agents if they get past the nonspecific defenses of the body. When the body is infected by a disease-causing agent, the immune system recognizes marker substances called antigens on the surface of pathogens. Antigens trigger the specific defenses of the immune system to produce antibodies specifically formed to fight that particular pathogen.

Antibodies are special proteins produced by B-lymphocyte cells of the immune system. Antibodies fight infection by binding to the antigens and clustering them into large groups. The clustering of antibodies and antigens is called agglutination. Agglutination prevents pathogens from infecting healthy cells and attracts phagocytes, which engulf and destroy the whole cluster of cells. This process is known as the primary immune response.

Once the body has been infected by a pathogen, the B- and T-lymphocytes that produced the primary immune response are capable of producing large numbers of anti-bodies against the same pathogen should it reappear in the body. This secondary immune response is so powerful and quick that the disease does not have a chance to develop in the body—it is immune to that disease.

In this section you also learned about three different types of immunity. A person has active immunity if he or she has been vaccinated or has recovered from the disease. Because the body produces antibodies in response to the vaccine or pathogens, it will always be able to mount a secondary immune response and actively provide the body with immunity from that particular pathogen. Passive immunity is produced when antibodies produced by another organism are injected into the bloodstream. Passive immunity provides temporary protection against a disease. The third type of immunity is cell-mediated immunity. All the cells of the body are marked with a special protein so that the immune system can recognize them. Cell-mediated immunity causes special killer T-cells to attack and destroy cells that lack the body's marker.

### A Perfect Match: Antigens and Antibodies

**1.** What is an antigen? _____

_____

_____

**2.** What is an antibody? _____

_____

_____

**3.** What role do antigens play in the immune system? _____

_____

_____

**4.** Match the antigen with the correct antibody.

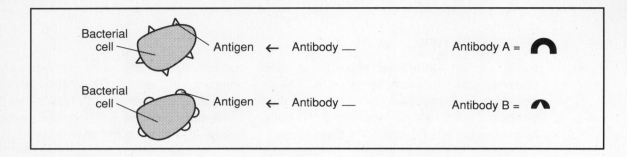

### The Immune Response: Sequencing Events

Use the numbers 1 through 5 to correctly sequence the events in the immune system response to an infection.

_____ B-lymphocytes produce antibodies to fight the infection.

_____ Agglutination causes antigens and antibodies to clump together.

_____ The immune system recognizes the antigen on the surface of the pathogen.

_____ The clumps of antigens and antibodies are engulfed and destroyed by phagocytes.

_____ Antibodies bind to the antigens on the pathogen.

### Comparing Types of Immunity

What type of immunity is described by each phrase?

**1.** Produced by an injection of antibodies: _____

**2.** Permanent immunity: _____

**3.** Produced when the body is injected with a weakened form of a pathogen:

_____

**4.** Killer T-cells destroy foreign cells: _____

**5.** Body makes its own antibodies in response to antigens: _____

**6.** Can cause rejection of transplanted organs: _____

### Concept Mapping

The construction of and theory behind concept mapping are discussed on pages vii–ix in the front of this Study Guide. Read those pages carefully. Then consider the concepts presented in Section 45–2 and how you would organize them into a concept map. Now look at the concept map for Chapter 45 on page 434. Notice that the concept map has been started for you. Add the key facts and concepts you feel are important for Section 45–2. When you have finished the chapter, you will have a completed concept map.

## Immune Disorders

*(pages 977–981)*

### SECTION REVIEW

The immune system provides the body with protection from a wide variety of diseases. But the immune system is not invincible! When the immune system malfunctions or is itself invaded by pathogens, the results are uncomfortable and sometimes deadly.

In this section you learned about allergies. Allergies are overreactions by the immune system to antigens. The symptoms that many allergy sufferers experience—runny nose and eyes and sneezing—are produced when antigens from plant pollen, dust, molds, or animal fur bind to mast cells located in the nasal passages. When the antigens bind to the mast cells, the mast cells release histamines, and it is the histamines that are responsible for the symptoms of allergies. Asthma is a serious allergy that causes the smooth muscles around the breathing passages to contract, making breathing very difficult.

The immune system is able to defend the body because it can recognize the cells that belong to the body and those that do not. An autoimmune disease results when the immune system, for some reason, cannot tell the difference between the body's cells and foreign cells and attacks the cells of the body. Examples of autoimmune diseases include rheumatic fever, rheumatoid arthritis, juvenile-onset diabetes, and multiple sclerosis.

You also learned about the deadly disease AIDS. AIDS is caused when the HIV virus infects and kills the helper T-cells of the immune system. The body recognizes the infection and produces antibodies against the HIV virus. But the antibodies are unable to destroy HIV because it is located within the cells of the immune system. Gradually, HIV kills off most helper T-cells and the body is unable to fight any infection. Repeated infections weaken the body, and eventually death results. HIV is present in the body's secretions and in blood, and it is through the exchange of these fluids that HIV and AIDS are spread. Any situations in which blood or body fluids are exchanged can result in the spread of AIDS: sexual intercourse, the sharing of needles by drug users, or from mother to fetus through the placenta. As yet, there is no cure for AIDS. To prevent AIDS, prevent exposure to the HIV virus!

### Disorders of the Immune System: Building Vocabulary Skills

Identify the disorder of the immune system described by each group of terms.

**1.** Helper T-cells, HIV, no cure: _____

_____

**2.** Attacks body's own cells, rheumatic fever: _____

_____

**3.** Mast cells, histamines, asthma: _____

_____

▨ **AIDS: Using the Main Ideas**

**1.** What part of the immune system is affected by HIV virus? _____

_____

_____

_____

**2.** Why are the HIV antibodies produced by the body useless against the virus?

_____

_____

_____

**3.** How does AIDS affect the body's ability to fight infection? _____

_____

_____

_____

**4.** Check the statements that describe a situation in which the AIDS virus could be spread.

_____ Holding hands with someone

_____ Sexual intercourse

_____ Coughing or sneezing

_____ Intravenous drug use

_____ Through the placenta from mother to unborn child

▨ **Concept Mapping**

The construction of and theory behind concept mapping are discussed on pages vii–ix in the front of this Study Guide. Read those pages carefully. Then consider the concepts presented in Section 45–3 and how you would organize them into a concept map. Now look at the concept map for Chapter 45 on page 434. Notice that the concept map has been started for you. Add the key facts and concepts you feel are important for Section 45–3. When you have finished the chapter, you will have a completed concept map.

## Using the Writing Process                                     *Chapter 45*

Use your writing skills and imagination to respond to the following writing assignment. You will probably need an additional sheet of paper to complete your response.

Pretend you are a war correspondent in the war between phagocytes and white blood cells. Write a news article in which you describe the different battles that take place in the body on a typical day. Make sure you indicate the nature of your newspaper. For example, your article might be very different if written for the *Plasma Gazette* as opposed to the *Viral Daily*.

_____

_____

_____

_____

_____

_____

_____

_____

_____

_____

_____

_____

_____

_____

_____

_____

_____

_____

_____

_____

_____

## Concept Mapping

The concept map below has been started for you. Add the key facts and concepts for each section of the chapter to this partial concept map. When you are done, you will have a concept map for the entire chapter.

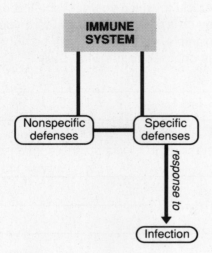

S T U D Y
G U I D E

CHAPTER
*Drugs, Alcohol, and Tobacco* **46**

### SECTION REVIEW

In this section you learned that a drug is any substance that causes a change in the body. Whether the substance is a legal or an illegal drug, it will produce some type of change in the body. It may kill disease-causing microorganisms, affect one of the body's systems, or change the user's behavior in some way. You also learned that using a drug in a way that doctors would not approve of is called drug abuse. Drug abuse is a major concern in our society because it causes psychological and physical problems for those involved in it. Continued abuse of a drug can lead to drug addiction, the uncontrollable craving for the drug. Drug addiction can be physical—the body has become dependent on a supply of the drug—or psychological—the abuser "thinks" the drug is needed in order for the person to function.

In this section you also learned about the effects that some commonly abused drugs have on the body. The most widely abused illegal drug is marijuana. The substance in marijuana that affects the body is called THC, and it produces a temporary feeling of euphoria and disorientation. Long-term abuse of marijuana can cause lung damage, loss of memory, loss of the ability to concentrate, reduced levels of testosterone in males, and psychological addiction.

Hallucinogens are drugs that affect the user's view of reality by interfering with the flow of nerve impulses in the brain. Examples of hallucinogens are LSD, or "acid," and PCP, or "angel dust."

Stimulants are drugs that speed up the actions of the nervous system by resembling the body's neurotransmitters, the substances that pass nerve impulses from neuron to neuron. For a short period of time, users feel full of strength and energy, then fatigue and depression return. Long-time abusers of stimulants may experience hallucinations, circulatory problems, and psychological problems.

Depressants have the opposite effect on the body as compared to stimulants. Depressants slow down the actions of the nervous system. When abused, depressants lead to physical addiction and require the help of a doctor for withdrawal.

A dangerous drug that can be smoked, sniffed, or injected is cocaine. Cocaine fools the brain into releasing the neurotransmitter dopamine. Dopamine produces a strong feeling of pleasure and satisfaction to which the cocaine abuser quickly becomes psychologically addicted. In addition to producing a sense of well-being, cocaine physically affects the body by increasing the user's heart rate and blood pressure. These physical effects can cause fatal heart attacks, even in young and healthy users! The most addictive form of cocaine is crack, which is smoked. Crack can lead to psychological dependence even after only a few doses!

Opiates are a group of drugs produced from the opium poppy. Opiates such as morphine are legal drugs used by doctors to relieve pain. Heroin, an illegal opiate, produces a sleepy feeling of well-being when it is injected into the bloodstream. Abuse of heroin leads to psychological and physical addiction.

▓ **Drug Abuse: Applying Vocabulary Skills**

Using the list below, select the term best illustrated in each situation.

Drug abuse                    Physical dependence
Drug addiction                Drug
Psychological dependence      Withdrawal

**1.** Joe had a headache, so he took an aspirin to relieve the pain. About 30 minutes after he

took the aspirin, Joe's head felt much better. _____

_____

_____

**2.** Mary had an uncontrollable urge for another dose of crack. _____

_____

_____

**3.** George had a terrible cold and had been taking some over-the-counter cold medicine.
The directions called for two tablets every four hours, but George felt so bad Monday

morning that he took three tablets instead of two. _____

_____

_____

**4.** Cathy had been using heroin for three years, and every time she tried to quit, she

experienced severe pain, nausea, chills, and fever. _____

_____

_____

**5.** Ben thought he had to have a marijuana cigarette to make it through the day.

_____

_____

_____

**6.** Jean had been taking sleeping pills in order to sleep at night. When she tried to quit, she

realized that her body had become dependent on the drug. _____

_____

_____

## Effects of Drugs: Using the Main Ideas

Complete the chart.

| Drug Type | Example | Effects on Body | Type(s) of Dependence |
|---|---|---|---|
| | Opium, morphine, and heroin | | |
| | | Affect user's view of reality | |
| Depressants | | | |
| | | | Psychological |
| | | Speed up the actions of the nervous system | |
| Marijuana | — | | |

## Concept Mapping

The construction of and theory behind concept mapping are discussed on pages vii–ix in the front of this Study Guide. Read those pages carefully. Then consider the concepts presented in Section 46–1 and how you would organize them into a concept map. Now look at the concept map for Chapter 46 on page 443. Notice that the concept map has been started for you. Add the key facts and concepts you feel are important for Section 46–1. When you have finished the chapter, you will have a completed concept map.

## Section 46–2  Alcohol

### SECTION REVIEW

In this section you learned about the oldest drug known to humans, the most dangerous drug, and the most abused drug in the world: alcohol! Alcohol is able to enter the bloodstream immediately because it is a small molecule that easily passes through cell membranes. Once in the bloodstream, alcohol has an almost instant effect on the nervous system. Because alcohol is a depressant, it slows down the rate at which the nervous system functions. After the first drink, a person may feel relaxed and confident. As more alcohol is consumed, a person's reaction time is longer, coordination is poor, and judgment is impaired. These effects make it dangerous to perform many activities after drinking—especially driving!

You also learned that as with any other drug, abuse can lead to serious physical and psychological problems. People who cannot function properly without satisfying their uncontrollable need for alcohol are alcoholics and suffer from the disease called alcoholism. Long-term effects of alcohol abuse include damage to the digestive system, destruction of the neurons in the brain, and the formation of scar tissue in the liver. Alcohol can also cause birth defects in the developing fetus if the mother drinks on a regular basis while she is pregnant. In addition to the physical problems caused by alcoholism, the psychological dependence often requires support groups that can provide help for the alcoholic to quit drinking and support for the family of the alcoholic.

### Alcohol: Using the Main Ideas

**1.** Why is alcohol the most dangerous and widely abused drug in the world?

_____

_____

_____

**2.** A person who abuses alcohol is called an _____

and suffers from the disease called _____

**3.** What form of alcohol is found in alcoholic beverages? What is its chemical formula?

_____

_____

**4.** What is fetal alcohol syndrome? How can it be prevented? _____

_____

_____

_____

_____

_____

### Effects of Alcohol: Interpreting Diagrams

Complete the diagram by explaining the effect or effects that alcohol can have on each of the body parts shown.

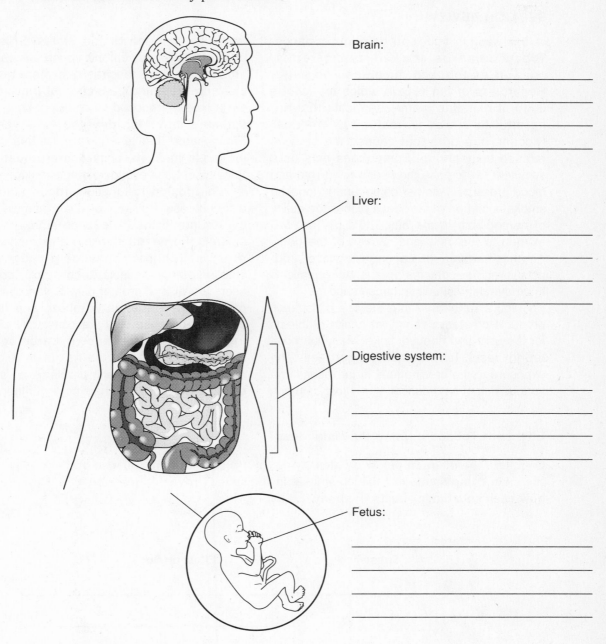

Brain:

_____

_____

_____

_____

Liver:

_____

_____

_____

_____

Digestive system:

_____

_____

_____

_____

Fetus:

_____

_____

_____

_____

### Concept Mapping

The construction of and theory behind concept mapping are discussed on pages vii–ix in the front of this Study Guide. Read those pages carefully. Then consider the concepts presented in Section 46–2 and how you would organize them into a concept map. Now look at the concept map for Chapter 46 on page 443. Notice that the concept map has been started for you. Add the key facts and concepts you feel are important for Section 46–2. When you have finished the chapter, you will have a completed concept map.

## Tobacco

### SECTION REVIEW

In this section you learned about tobacco. Tobacco can be smoked as cigarettes, chewed as tobacco, or dipped as snuff. Regardless of the way in which tobacco is used, it contains many chemicals that are harmful to the body, including nicotine. Nicotine is a drug that causes the body to release epinephrine. Epinephrine acts as a stimulant to increase the user's heart rate and blood pressure. Another compound in tobacco smoke is carbon monoxide; it blocks the ability of hemoglobin in the blood to carry needed oxygen to the cells and tissues of the body. Another harmful substance in tobacco products is tar. Tar contains a variety of chemicals that have been shown to cause cancer.

You also learned that the use of tobacco products creates a variety of health problems for the user and nonuser alike. Many lung disorders result from smoking cigarettes. The exposure to the carcinogens in tar results in a serious risk of lung cancer—the most common fatal cancer in the United States! Smokers may also develop a chronic cough and chronic bronchitis as the respiratory system tries to clear its breathing passages of the foreign substances introduced by smoking. Long-term smokers may also develop emphysema. Emphysema results when the tissues of the lung lose their elasticity and are unable to force air out of the lungs during exhalation.

In addition to lung disorders, smokers also increase their risk of heart disease. The carbon monoxide in tobacco smoke slowly poisons the heart. Smoking also makes the heart work harder because it causes the smoker's blood vessels to constrict. Tobacco users are also at greater risk for various types of cancers—particularly mouth and throat cancers—because they are constantly exposing themselves to known carcinogens. Nonsmokers also run the risk of developing many of the same health problems as smokers if they are inhaling the smoke of others.

### Effects of Tobacco: Using the Main Ideas

Complete the diagram below by identifying the primary drug in tobacco and the main components of tobacco smoke. In the space provided, describe how each substance affects the body.

| Substance | Effect on Body |
|---|---|
| _____ : | _____ |
|  | _____ |
| _____ : | _____ |
|  | _____ |
| _____ : | _____ |
|  | _____ |

**Tobacco and Disease: Applying Concepts**

**1.** How is each of the labeled structures in the respiratory system affected by smoking?

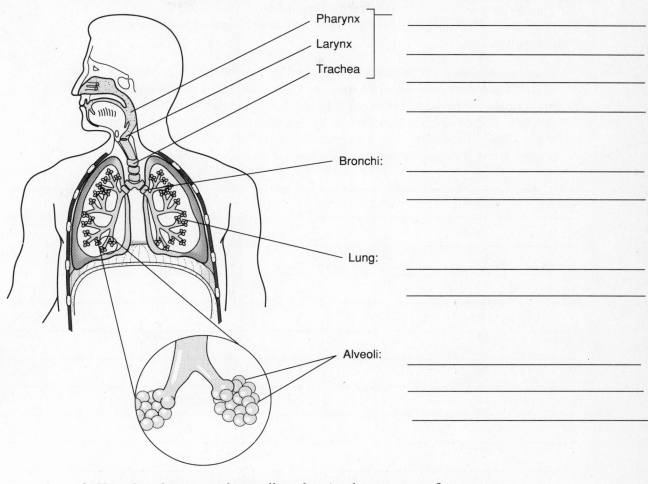

Pharynx ⎤
Larynx  ⎬    _____
Trachea ⎦    _____
             _____
             _____

Bronchi:     _____
             _____

Lung:        _____
             _____

Alveoli:     _____
             _____
             _____

**2.** How do tobacco products affect the circulatory system?

_____

_____

_____

_____

_____

**Concept Mapping**

The construction of and theory behind concept mapping are discussed on pages vii–ix in the front of this Study Guide. Read those pages carefully. Then consider the concepts presented in Section 46–3 and how you would organize them into a concept map. Now look at the concept map for Chapter 46 on page 443. Notice that the concept map has been started for you. Add the key facts and concepts you feel are important for Section 46–3. When you have finished the chapter, you will have a completed concept map.

## Using the Writing Process                    *Chapter 46*

Use your writing skills and imagination to respond to the following writing assignment. You will probably need an additional sheet of paper to complete your response.

Drug dealers have started targeting elementary school students. Some of these students are easy prey because they are not aware of the dangers of drugs. An educational network has decided to use puppet shows to combat this problem. You have been approached by this television network to write one of these shows. Provide a written outline, a script, or a storyboard of the first puppet show. If possible, perform the show for your classmates.

_____

_____

_____

_____

_____

_____

_____

_____

_____

_____

_____

_____

_____

_____

_____

_____

_____

_____

_____

_____

## Concept Mapping

The concept map below has been started for you. Add the key facts and concepts for each section of the chapter to this partial concept map. When you are done, you will have a concept map for the entire chapter.

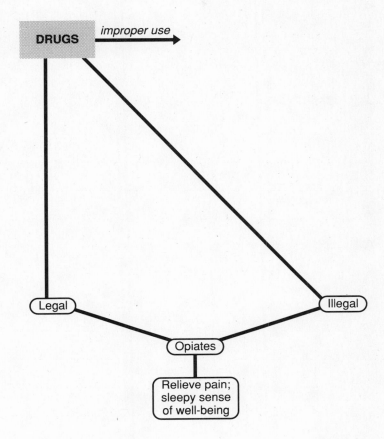

S T U D Y
G U I D E

| Section 47–1 | Earth: A Living Planet | (pages 1007–1010) |

### SECTION REVIEW

In this section you were introduced to the branch of science called ecology. Ecology is the study of the interactions of organisms with one another and with their physical surroundings.

You learned that the Earth is a single living system that has been named the biosphere, or living globe. You also learned that the biosphere is divided into a variety of smaller units called ecosystems. Ecosystems consist of all the living and nonliving factors that surround organisms and affect their way of life. Some examples of ecosystems are pond ecosystems and forest ecosystems.

As you continued studying this section you learned that environments of organisms are constantly changing in a process called ecological succession. In ecological succession, an existing community is gradually replaced by another community. In many cases ecological succession leads to a relatively stable collection of plants and animals called a climax community.

### Understanding Relationships: Building Vocabulary Skills

In the space provided, explain how the paired terms are related to each other.

**1.** biosphere, ecosystem: _____

_____

_____

_____

_____

**2.** ecology, biosphere: _____

_____

_____

_____

_____

**3.** biotic factors, abiotic factors: _____

_____

_____

_____

_____

4. ecological succession, climax community: _____

_____

_____

_____

5. community, ecosystem: _____

_____

_____

_____

### Ecological Succession: Understanding the Main Ideas

The pictures in the accompanying illustration show four stages in the development of an ecological community on the site of an abandoned farm. These stages are not in order.

**A.** Small fast-growing trees such as poplars and sumacs

**B.** Grasses and small wildflowers such as asters and dandelions

**C.** Large trees such as maples, oaks, and birches

**D.** Weeds and small shrubs

1. Which picture (A, B, C, or D) shows the first stage of succession?_____

   The second? _____

   The third? _____

   The fourth? _____

2. What kinds of plants might be associated with the climax community in this area?

   _____

   _____

   _____

   _____

   _____

   _____

   _____

3. Suppose a site similar to this one was mowed twice a year. What kinds of plants would

   you expect to find on such a site? Explain._____

   _____

   _____

   _____

   _____

   _____

   _____

   _____

### Concept Mapping

The construction of and theory behind concept mapping are discussed on pages vii–ix in the front of this Study Guide. Read those pages carefully. Then consider the concepts presented in Section 47–1 and how you would organize them into a concept map. Now look at the concept map for Chapter 47 on page 456. Notice that the concept map has been started for you. Add the key facts and concepts you feel are important for Section 47–1. When you have finished the chapter, you will have a completed concept map.

## SECTION REVIEW

A biome is an environment that has a characteristic climax community. The Earth is made up of two main types of biomes: land biomes and aquatic biomes. In this section you learned about the various types of land biomes found on Earth. You learned that the major land biomes are the tundra, taiga, temperate deciduous forest, grassland, tropical rain forest, and desert.

Each land biome is characterized by a particular type of climate and physical features. For example, the tundra is very cold and dry, whereas the tropical rain forest is warm and wet. Each land biome is also characterized by particular types of plant and animal life. For example, the taiga is covered with coniferous forests, whereas the grasslands are covered with grasses and small leafy plants.

The various land biomes are found in certain parts of the world. Recall that the polar regions of North America, Europe, and Asia are covered by tundra. South of the tundra is the taiga. Tropical rain forests tend to be found near the equator. Grasslands usually cover the interior areas of continents.

### Recalling Information: Building Vocabulary Skills

Read each of the following descriptions carefully. Decide which land biome best fits each description. Then write the name of the biome you have selected on the blank line next to the description.

_____ **1.** Coniferous trees such as pine, spruce, and fir are the dominant plants.

_____ **2.** Trees display brilliant colors in autumn.

_____ **3.** Soil is characterized by permafrost.

_____ **4.** There is less than 25 centimeters of rainfall a year.

_____ **5.** Wheat, corn, and other grains are heavily farmed in this biome in the United States and the Soviet Union.

_____ **6.** This warm biome receives 200 to 400 centimeters of rainfall every year.

_____ **7.** Typical inhabitants include grizzly bears, moose, and wolves.

_____ **8.** Periodic fires and the grazing of large animals prevent the growth of forests.

_____ **9.** This is the northernmost land biome.

### Land Biomes of the World: Using the Main Ideas

The map below shows where each type of land biome is found on Earth.
Complete the key by writing the name of the correct biome next to each
type of shading. Then answer the questions that follow.

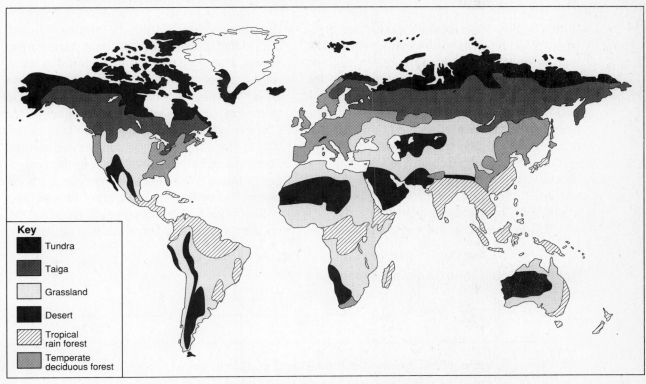

**Key**
Tundra
Taiga
Grassland
Desert
Tropical rain forest
Temperate deciduous forest

1. Which biomes are found in the mainland United States? _____

   _____

2. Including Alaska and Hawaii, which additional biomes are found in the United States?

   _____

3. Which biome makes up the greatest portions of Canada and the Soviet Union?

   _____

4. In which parts of the world are tropical rain forests found? _____

   _____

### Concept Mapping

The construction of and theory behind concept mapping are discussed on
pages vii–ix in the front of this Study Guide. Read those pages carefully. Then
consider the concepts presented in Section 47–2 and how you would organize
them into a concept map. Now look at the concept map for Chapter 47 on
page 456. Notice that the concept map has been started for you. Add the key
facts and concepts you feel are important for Section 47–2. When you have
finished the chapter, you will have a completed concept map.

## SECTION REVIEW

In this section you learned about the three types of aquatic biomes: freshwater biomes, marine biomes, and estuaries. You learned that rivers, streams, and lakes make up the freshwater biomes of the Earth, whereas the habitats of the ocean make up the marine biomes. You discovered that some of the abiotic factors that affect the kinds of organisms found in the aquatic biomes are light intensity, amounts of oxygen and carbon dioxide dissolved in the water, and the availability of organic and inorganic nutrients.

You learned that the marine biomes are divided into ecologically distinct zones depending on depth and the distance from shore. These zones are the intertidal zone, neritic zone, open-sea zone, and deep-sea zone. You also learned about the importance of the photic zone, which is the region of the ocean in which photosynthesis can take place.

In the last part of the section you learned about estuaries. Estuaries exist at the boundary between a freshwater biome and a marine biome. These are shallow areas that support a variety of life forms. Some examples of estuaries are salt marshes, swamps, and lagoons.

### Applying Definitions: Building Vocabulary Skills

**1.** What are aquatic biomes? _____

_____

**2.** Compare and contrast freshwater biomes and marine biomes.

_____

_____

_____

_____

_____

**3.** Explain the relationship that exists between freshwater biomes, marine biomes, and

estuaries. _____

_____

_____

**4.** In what type of biome is the photic zone found? Why is it important?

_____

_____

_____

_____

■ **Marine Biomes: Applying the Main Ideas**

**1.** Label the zones on the diagram of the ocean.

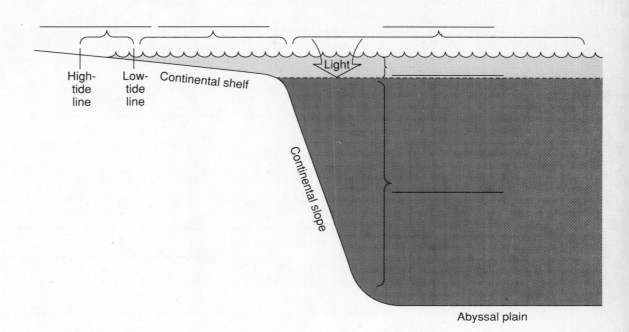

**2.** In your own words, describe each zone in the space provided.

Photic zone: _____

_____

_____

_____

_____

Intertidal zone: _____

_____

_____

_____

_____

Neritic zone: _____

_____

_____

_____

_____

Open-sea zone: _____

_____

_____

_____

_____

_____

Deep-sea zone: _____

_____

_____

_____

_____

_____

**3.** Explain why photosynthesis cannot occur in the deep-sea zone.

_____

_____

_____

_____

_____

**4.** How deep is the photic zone?_____

_____

_____

_____

### Concept Mapping

The construction of and theory behind concept mapping are discussed on pages vii–ix in the front of this Study Guide. Read those pages carefully. Then consider the concepts presented in Section 47–3 and how you would organize them into a concept map. Now look at the concept map for Chapter 47 on page 456. Notice that the concept map has been started for you. Add the key facts and concepts you feel are important for Section 47–3. When you have finished the chapter, you will have a completed concept map.

**Section 47–4**  **Energy and Nutrients: Building the Web of Life**  *(pages 1021–1027)*

### SECTION REVIEW

In the first part of this section you learned that energy flows through an ecosystem from the sun to producers and then to consumers. You also were introduced to terms used to describe the energy relationships within an ecosystem. Some of the more important terms and concepts to remember include producer, consumer, decomposer, trophic level, and ecological pyramid.

Although energy moves in a one-way direction through an ecosystem, nutrients are recycled. In the middle part of this section you read about the major biogeochemical cycles that move nutrients through the biosphere. These include the water cycle, the nitrogen cycle, the carbon cycle, and the oxygen cycle. You may recall that a nutrient that is in short supply may be a limiting factor.

In the last part of this section you learned that organisms are tied together by complicated networks of feeding relationships. These networks are known as food webs. A single isolated strand of a food web is called a food chain.

### Understanding Key Terms: Building Vocabulary Skills

The terms in each of the following groups are related to one another. For each group, define the terms and state the relationships that exist among the terms.

**1.** producers, consumers, decomposers: _____

_____

_____

_____

**2.** trophic level, ecological pyramids: _____

_____

_____

_____

**3.** biogeochemical cycles, water cycle, nitrogen cycle, carbon cycle, oxygen cycle:

_____

_____

_____

**4.** nitrogen cycle, nitrogen fixation, denitrification: _____

_____

_____

_____

### Nutrients and Energy: Understanding the Main Ideas

Using your own words, explain how the movement of nutrients in an ecosystem is different

from the movement of energy in an ecosystem. _____

_____

_____

_____

### Where Are You in the Food Chain?

Imagine that you have just eaten a tuna sandwich, an apple, and a glass of milk. In the space provided, draw a possible food chain for each item listed. Then indicate which order you are in each chain.

**1.** Tuna: _____

_____

_____

_____

Order: _____

**2.** Apple: _____

_____

_____

_____

Order: _____

**3.** Milk: _____

_____

_____

_____

Order: _____

### Concept Mapping

The construction of and theory behind concept mapping are discussed on pages vii–ix in the front of this Study Guide. Read those pages carefully. Then consider the concepts presented in Section 47–4 and how you would organize them into a concept map. Now look at the concept map for Chapter 47 on page 456. Notice that the concept map has been started for you. Add the key facts and concepts you feel are important for Section 47–4. When you have finished the chapter, you will have a completed concept map.

## Using the Writing Process

Use your writing skills and imagination to respond to the following writing assignment. You will probably need an additional sheet of paper to complete your response

Imagine that a climax community wishes to honor the communities that came before it and made its existence possible. The members of the climax community decide to build a memorial and hold a dedication ceremony. Design this memorial. Then write a dedication speech in which you praise the accomplishments of the previous communities. In your speech you should also explain any symbolism in the memorial design.

_____

_____

_____

_____

_____

_____

_____

_____

_____

_____

_____

_____

_____

_____

_____

_____

_____

_____

_____

_____

_____

## Concept Mapping

The concept map below has been started for you. Add the key facts and concepts for each section of the chapter to this partial concept map. When you are done, you will have a concept map for the entire chapter.

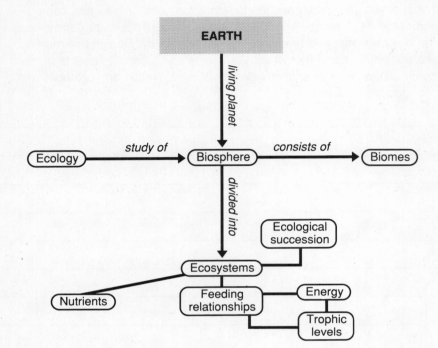

S T U D Y
G U I D E

CHAPTER **48**
*Population and Communities*

**Section 48–1**   **Population Growth**

*(pages 1033–1035)*

### SECTION REVIEW

When provided with ideal conditions for growth and reproduction, almost any organism will experience a rapid increase in its population. If nothing stops the population from growing, it will experience exponential growth. However, exponential growth does not continue in natural populations for long. Most populations go through a number of growth phases, producing a pattern that is known as logistic growth.

In this section you learned that an environment has a carrying capacity for a particular species. This means that the environment can support a population of a certain size and no more. Factors such as the amount of food and space as well as competition among individuals determine the carrying capacity. Once a population reaches the carrying capacity of its environment, its average growth rate will be zero.

### Interpreting Graphs: Using the Main Ideas

The graphs below represent two different types of population growth. Use the graphs to answer the questions that follow.

_____    _____

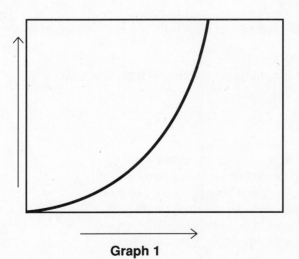

**Graph 1**

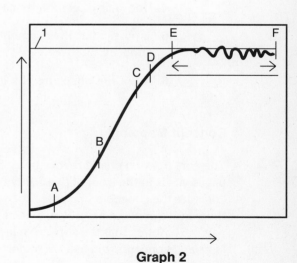

**Graph 2**

1. On the blank lines above the graphs, give each graph a title based on the type of population growth it depicts.

2. On each graph, what does the horizontal axis represent? _____

   What does the vertical axis represent? _____

   Label the vertical and horizontal axes on both graphs.

**3.** On Graph 2, what does the area between points E and F represent? _____

_____

Label this area.

**4.** On Graph 2, what does line l represent? _____
Label line l.

**5.** Describe in your own words what is happening to the population in Graph 1.

_____

_____

_____

_____

### Applying Definitions: Building Vocabulary Skills

Each of the statements below refers to exponential population growth or logistic population growth. In the blank next to each statement, place an E if the statement refers to exponential growth and an L if the statement refers to logistic growth.

_____ **1.** A lack of food prevents a certain population from growing any further.

_____ **2.** In the steady state, the average growth rate is zero.

_____ **3.** The larger the population gets, the faster it grows.

_____ **4.** One pair of elephants could produce 19 million offspring in less than 750 years.

_____ **5.** A particular environment is limited to a certain number of rabbits that it can support.

_____ **6.** All of the offspring of a given population survive and reproduce.

### Concept Mapping

The construction of and theory behind concept mapping are discussed on pages vii–ix in the front of this Study Guide. Read those pages carefully. Then consider the concepts presented in Section 48–1 and how you would organize them into a concept map. Now look at the concept map for Chapter 48 on page 464. Notice that the concept map has been started for you. Add the key facts and concepts you feel are important for Section 48–1. When you have finished the chapter, you will have a completed concept map.

**Section 48-2** | **Factors That Control Population Growth** | *(pages 1035-1040)*

### SECTION REVIEW

In this section you learned about the factors that control population growth. These factors enable a population to maintain levels that are between extinction and overpopulation.

The factors that control population can be classified as either density-dependent limiting factors or density-independent limiting factors. Density-dependent limiting factors are those factors that usually operate only when a population is large. These include competition, predation, parasitism, and overcrowding and stress. Density-independent limiting factors affect a population regardless of size. These include natural occurrences such as rainstorms, frost, wind, and temperature changes.

In the last part of this section you read about human population growth. You discovered that the world's human population has been growing exponentially for about the last 500 years, but its growth has slowed down in areas such as the United States and Europe.

### Understanding Definitions: Building Vocabulary Skills

Each of the statements below describes a situation that affects population growth. Some of the statements describe density-dependent factors, whereas others describe density-independent factors. In the blank before each item, place a D if the statement describes a density-dependent factor and an I if the statement describes a density-independent factor.

_____ **1.** A severe frost wipes out 50 percent of the citrus crop in southern Florida.

_____ **2.** Since snakes prey on frogs, an increase in the frog population causes an increase in the snake population.

_____ **3.** Due to severe overcrowding in an Asian village, many children do not survive to reach adulthood.

_____ **4.** The eruption of Mt. St. Helens destroys most of the wildlife in the immediate vicinity of the volcano.

_____ **5.** Off the coast of Peru, many fish die due to a change in the winds and the movement of ocean currents.

_____ **6.** Two animals attempt to occupy the same niche. The more aggressive animal survives, and the other does not.

_____ **7.** Because rabbits in Australia have no natural enemies, their population increases exponentially.

_____ **8.** Travelers who venture into a crowded African village become infected with a disease caused by parasites.

_____ **9.** Fish on a coral reef stake out their territory and chase away any younger fish that try to live there.

_____ **10.** Due to stress, large numbers of female mice miscarry their young and fail to reproduce.

▨ **Interpreting a Graph: Using the Main Ideas**

The graph below shows a predator-prey relationship. Use the graph to answer the questions that follow.

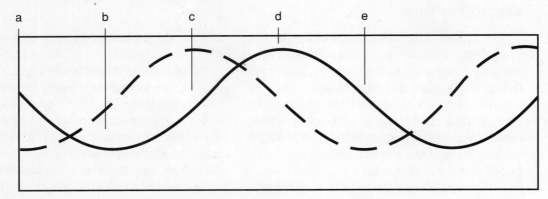

Key
— — — Prey
————— Predator

1. What is happening to the predator population between points c and d on the graph?

_____

2. What is causing this to happen? _____

_____

3. What is happening to the predator population at point e? Why is this happening?

_____

4. What is happening to the prey population at point b? Why is this happening?

_____

5. Suppose that human activity were to wipe out a large number of predators between points c and d. Predict what would happen to the prey population between points

c and e. _____

_____

▨ **Concept Mapping**

The construction of and theory behind concept mapping are discussed on pages vii–ix in the front of this Study Guide. Read those pages carefully. Then consider the concepts presented in Section 48–2 and how you would organize them into a concept map. Now look at the concept map for Chapter 48 on page 464. Notice that the concept map has been started for you. Add the key facts and concepts you feel are important for Section 48–2. When you have finished the chapter, you will have a completed concept map.

| Section 48–3 | **Interactions Within and Between Communities** | *(pages 1040–1043)* |

## SECTION REVIEW

In this section you learned about the ways in which populations within communities interact with one another. You also learned about how ecosystems interact with one another.

A community consists of all the populations of organisms living in a given area. You discovered that an important set of relationships called symbiosis occurs within a community. Two types of symbiosis that were discussed in this section are commensalism and mutualism. You will recall from the previous section that parasitism is also a form of symbiosis.

You learned that ecosystems do not exist in isolation from one another. Nearly every ecosystem is connected either directly or indirectly with several other ecosystems. It is important to keep this in mind, for it is one of the reasons why humans must be careful about how they dispose of wastes in the environment.

### Understanding Definitions: Building Vocabulary Skills

Each item below describes a population, a community, or neither. In the blank before each item, write P if the item describes a population; C if the item describes a community; and N if the item describes neither.

_____ **1.** All the plants and animals found in a pond ecosystem

_____ **2.** All the people, plants, and animals living in Toledo, Ohio

_____ **3.** All the finback whales alive in the world today

_____ **4.** A particular species of cactus found in a desert ecosystem

_____ **5.** A particular species of bass found in a lake in upstate New York

_____ **6.** All the squirrels, chipmunks, and gophers found in a forest ecosystem

_____ **7.** All the humans living on the north shore of Echo Lake

_____ **8.** All the living things found at the bottom of the ocean in an area of the deep-sea zone

_____ **9.** All the oysters and clams found in the ocean off the coast of New England

### Symbiosis: Understanding the Main Idea

Each of the following situations describes a form of symbiosis. After each situation, write the type of symbiosis that is being described. Then explain why this form of symbiosis applies to this situation.

parasitism          commensalism          mutualism

***Situation 1:*** A flowering plant cannot pollinate another plant unless pollen is transported between the plants. While gathering nectar from the plant's flowers, a bee is lightly dusted with pollen.

The bee then transports the pollen as it moves from one flower to another. This enables the flowering plant to reproduce.

_____

_____

_____

***Situation 2:*** A liver fluke enters the human digestive system on a piece of beef. The fluke derives nourishment from the human; the human is seriously weakened by the presence of the fluke.

_____

_____

_____

***Situation 3:*** A fungus that cannot make its own food absorbs sugars and other nutrients from plant roots; the fungus also absorbs from the soil certain minerals that can be used by the plant and which the plant has difficulty obtaining by itself.

_____

_____

_____

***Situation 4:*** Tall trees provide birds with a place to nest that offers them protection against many kinds of predators.

_____

_____

***Situation 5:*** A blind human is able to move about safely with the help of a guide dog. The human provides the dog with food, shelter, and medical care.

_____

_____

## Concept Mapping

The construction of and theory behind concept mapping are discussed on pages vii–ix in the front of this Study Guide. Read those pages carefully. Then consider the concepts presented in Section 48–3 and how you would organize them into a concept map. Now look at the concept map for Chapter 48 on page 464. Notice that the concept map has been started for you. Add the key facts and concepts you feel are important for Section 48–3. When you have finished the chapter, you will have a completed concept map.

**Using the Writing Process**                                   *Chapter 48*

Use your writing skills and imagination to respond to the following writing assignment. You will probably need an additional sheet of paper to complete your response.

All the members of an ecosystem decide that they need a legal document that will govern the relationships within that ecosystem. Representatives of the various populations within the ecosystem are called together for a constitutional convention. Describe the discussions that might take place as each population argues for the rules that will govern the ecosystem as well as the way the ecosystem interacts with other ecosystems. Or, if you prefer, draft a sample constitution that both governs and protects the rights of all the populations in the ecosystem.

_____

_____

_____

_____

_____

_____

_____

_____

_____

_____

_____

_____

_____

_____

_____

_____

_____

_____

_____

**463**

## Concept Mapping

The concept map has been started for you. Add the key facts and concepts for each section of the chapter to this partial concept map. When you are done, you will have a concept map for the entire chapter.

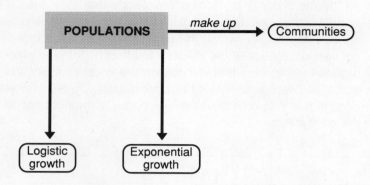

S T U D Y
G U I D E

CHAPTER
*People and the Biosphere* **49**

| Section 49-1 | **Human Population** | *(pages 1049-1052)* |

### SECTION REVIEW

In this section you learned how a growing human population has affected planet Earth. You also learned how modern lifestyles place a heavy demand on the environment.

In many parts of the world human population is growing rapidly. As a result, more and more land must be used for farming. This often means cutting down valuable forests. In industrialized nations, where population is not growing so rapidly, people's ways of life often have an adverse effect on the environment. For example, many consumer goods require the use of much energy and natural resources, and the production of these goods often causes pollution.

### Identifying Cause and Effect: Finding the Main Ideas

The diagram below shows how the development and growth of human populations have affected the environment. Use the diagram to answer the questions on the following page.

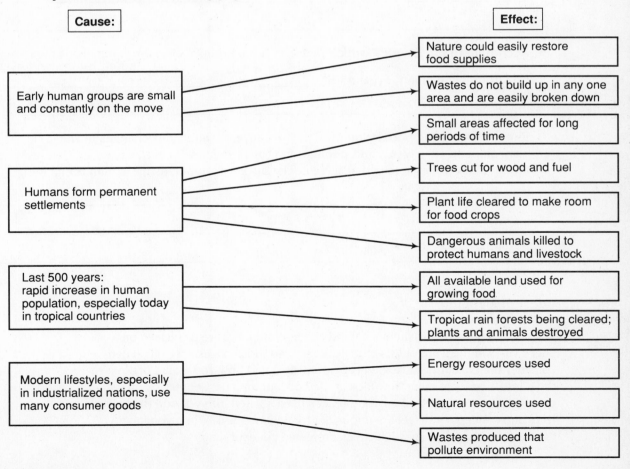

**Cause:**

Early human groups are small and constantly on the move

Humans form permanent settlements

Last 500 years:
rapid increase in human population, especially today in tropical countries

Modern lifestyles, especially in industrialized nations, use many consumer goods

**Effect:**

Nature could easily restore food supplies

Wastes do not build up in any one area and are easily broken down

Small areas affected for long periods of time

Trees cut for wood and fuel

Plant life cleared to make room for food crops

Dangerous animals killed to protect humans and livestock

All available land used for growing food

Tropical rain forests being cleared; plants and animals destroyed

Energy resources used

Natural resources used

Wastes produced that pollute environment

1. According to the diagram, what three factors in human population development have caused significant permanent changes in the environment?_____

_____

_____

_____

2. Why did the earliest human population groups have little lasting effect on the environment? _____

_____

3. In what ways has human population development caused the Earth's plant life to be disrupted? _____

_____

_____

4. In addition to those shown, can you think of any other human factors that have caused a disruption of plant life? _____

_____

_____

5. According to the diagram, which factor in human population development is most directly responsible for pollution?_____

_____

6. Rapidly increasing population growth is a cause of significant changes in the environment. However, rapid population growth has itself been caused by several factors. List three factors that have contributed to an increasing human population.

_____

_____

_____

**Concept Mapping**

The construction of and theory behind concept mapping are discussed on pages vii–ix in the front of this Study Guide. Read those pages carefully. Then consider the concepts presented in Section 49–1 and how you would organize them into a concept map. Now look at the concept map for Chapter 49 on page 476. Notice that the concept map has been started for you. Add the key facts and concepts you feel are important for Section 49–1. When you have finished the chapter, you will have a completed concept map.

| Section 49–2 | Pollution | (pages 1052–1063) |

## SECTION REVIEW

Pollution! How often do you hear that word? If you follow the news media carefully, you probably hear or read the word *pollution* almost every day. As you learned in this section, there are many ways in which human activities pollute the Earth—and pollution seems to be getting worse. Among the various kinds of pollution that threaten planet Earth are air pollution, freshwater pollution, ocean pollution, and biological magnification.

You learned that pollution stems largely from the production of wastes and the improper disposal of wastes. You also learned that materials released into the environment can be classified as either biodegradable or non-

biodegradable. Biodegradable materials can be broken down by microorganisms; nonbiodegradable materials cannot. As a result, nonbiodegradable materials remain pollutants in the environment for long periods of time.

Sometimes a seemingly minor pollutant enters the environment and causes much greater damage than might have been expected. This happens because the pollutant is concentrated as it is passed from one organism to another via food chains and food webs. This phenomenon is called biological magnification. Biological magnification occurs with many pesticides and industrial waste products.

### Understanding Definitions: Building Vocabulary Skills

In each item below, decide whether the underlined term is used correctly. In the blank after each item, write "yes" if the term is used correctly. If the term is not used correctly, write the correct term.

1. Materials that remain in the environment indefinitely are called <u>biodegradable</u>.

_____

2. When certain pollutants in the air combine with water vapor to produce drops of acid, the result is the <u>greenhouse effect</u>. _____

3. The <u>ozone layer</u> protects the Earth from harmful ultraviolet radiation from the sun.

_____

4. <u>Carbon dioxide</u> and other gases in the atmosphere absorb heat energy and form a kind of heat blanket around the Earth. _____

5. When factories dump <u>chemicals</u> into rivers and streams, thermal pollution results.

_____

6. When a pollutant is passed from one organism to another in a food chain, the concentration of the pollutant <u>decreases</u> in a process called biological magnification.

_____

7. Metals, glass, plastic, and radioactive wastes are <u>nonbiodegradable</u> materials.

_____

**8.** In a temperature inversion, a layer of cool <u>clean</u> air is trapped beneath a layer of warm

air. _____

### Pollution: Relating the Main Ideas

The diagram below shows many of the ways in which the biosphere
becomes polluted. Use the diagram to answer the questions that follow.

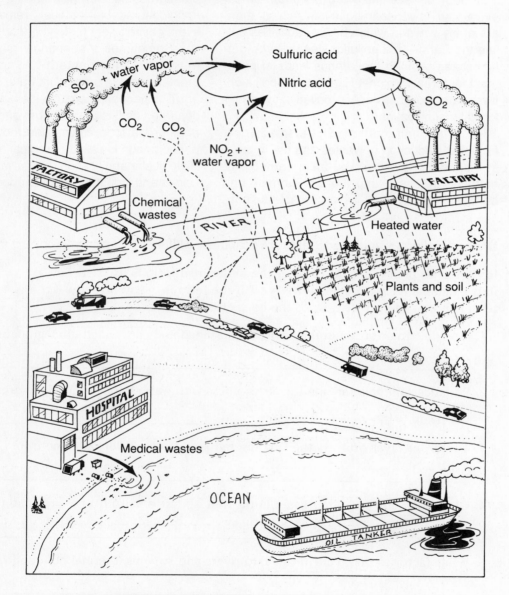

**1.** List the potential sources of pollution shown in the diagram. _____

_____

**2.** List three ways in which the river is being polluted. _____

_____

_____

© Prentice-Hall, Inc.

**3.** How might the plants and soil eventually be affected by these same pollutants?

_____

_____

_____

**4.** What two factors are shown as possible ocean pollutants? _____

_____

**5.** What kind of pollution—air, land, or water—is being caused by the automobiles? What will be the result of this pollution? _____

_____

_____

**6.** How does acid rain form? _____

_____

_____

_____

**7.** How does acid rain link together the pollution of air, land, and water?

_____

_____

_____

**8.** The diagram below shows a rather small area. How might neighboring areas or even distant areas be affected by these same sources of pollution? _____

_____

_____

_____

_____

### Concept Mapping

The construction of and theory behind concept mapping are discussed on pages vii–ix in the front of this Study Guide. Read those pages carefully. Then consider the concepts presented in Section 49–2 and how you would organize them into a concept map. Now look at the concept map for Chapter 49 on page 476. Notice that the concept map has been started for you. Add the key facts and concepts you feel are important for Section 49–2. When you have finished the chapter, you will have a completed concept map.

**The Fate of the Earth** *(pages 1063–1068)*

## SECTION REVIEW

In recent years it has become clear that the survival of humans and human society depends on the survival of other organisms. Unfortunately, human activities do not always promote the survival of other living things. In this section you learned that among the living things that are especially threatened by human activities are forests and endangered species.

Forests are essential to the health of the biosphere. When trees are cut down, many changes occur, and often these changes are harmful to the total environment. To try to repair some of the damage caused by the loss of forests, reforestation programs seek to replace trees that have been cut down by planting new trees.

Some animal and plant species are threatened with extinction because of reduced populations. Such species are called endangered species. Many groups and individuals have taken up the cause of saving endangered species. Gradually more and more people are becoming aware of the need to protect not only endangered species but all types of organisms.

### Understanding Definitions: Building Vocabulary Skills

1. What is reforestation? _____

_____

_____

_____

_____

2. Define the term *endangered species*. _____

_____

_____

_____

_____

3. List three factors that cause species to become endangered. _____

_____

_____

_____

_____

_____

_____

▓ **Tropical Rain Forests: Exploring the Main Ideas**

The first map below shows the tropical rain forests that exist in the world today. The second map shows what many experts predict will be the amount of tropical rain forests by the beginning of the next century. Use the maps to answer the questions that follow.

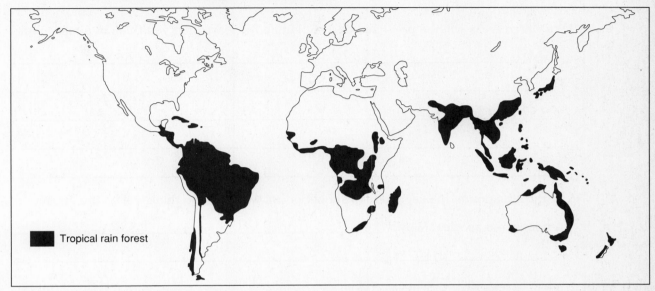

Tropical rain forest

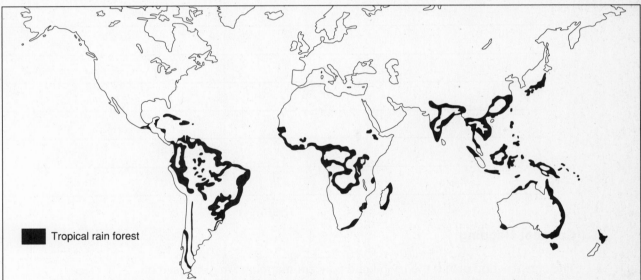

Tropical rain forest

**1.** In what parts of the world are tropical rain forests found? _____

_____

_____

_____

**2.** What continent has more than half of its area covered by rain forests?

_____

3. According to the maps, about how much of the world's tropical rain forests will be

   destroyed by the beginning of the next century? _____

   _____

   _____

4. In which areas will the predicted destruction be the least? Can you think of a reason for

   this? _____

   _____

   _____

   _____

   _____

5. If the situation in the second map should occur, what do you think will be the impact on

   endangered species? Why? _____

   _____

   _____

   _____

   _____

   _____

   _____

   _____

   _____

   _____

### Concept Mapping

The construction of and theory behind concept mapping are discussed on
pages vii–ix in the front of this Study Guide. Read those pages carefully. Then
consider the concepts presented in Section 49–3 and how you would organize
them into a concept map. Now look at the concept map for Chapter 49 on
page 476. Notice that the concept map has been started for you. Add the key
facts and concepts you feel are important for Section 49–3. When you have
finished the chapter, you will have a completed concept map.

## SECTION REVIEW

In this section you read about the need for people to preserve the biosphere through such efforts as conservation. You also read about the importance of protecting wild plants and animals, conserving natural resources, and finding alternative methods of waste disposal.

By the time you finished this section you discovered that protecting the environment often leads to difficult decisions. Many of these decisions center around economic factors. Certainly there is a price to be paid for a healthy biosphere. Fortunately, however, more people are becoming willing to pay the price and to make preserving the biosphere a major priority.

### Understanding Definitions: Building Vocabulary Skills

1. What is recycling? _____

_____

2. What types of items are suitable for recycling? _____

_____

_____

3. State at least two ways in which recycling helps to protect the environment.

_____

_____

_____

### The Future of the Biosphere: Interpreting the Main Ideas

The quotations below represent three viewpoints about people and the biosphere. Refer to the quotations as you answer the questions that follow.

**Quotation 1:**
*We don't inherit the earth from our ancestors, we borrow it from our children.*
— Old Pennsylvania Dutch expression

**Quotation 2:**
*We travel together on a little spaceship dependent on its vulnerable supplies of air and soil; all committed for our safety to its security and peace, preserved from annihilation only by the care, the work, and, I will say, the love we give our fragile craft.*
— Adlai Stevenson

**Quotation 3:**
*If we are not willing to change, we will disappear from the face of the globe, to be replaced by the insect.*
— Jacques Cousteau

1. Express in your own words the meaning of the first quotation.

   _____

   _____

2. How does the second quotation express the relationship of the people and the

   biosphere? _____

   _____

   _____

3. In the third quotation, what species appears to be in danger of extinction? Explain.

   _____

   _____

   _____

4. If you were going to make a poster urging people to be more concerned about the

   environment, which quotation would you choose? Why? _____

   _____

   _____

5. Make a drawing of your poster in the space below.

## Concept Mapping

The construction of and theory behind concept mapping are discussed on
pages vii–ix in the front of this Study Guide. Read those pages carefully. Then
consider the concepts presented in Section 49–4 and how you would organize
them into a concept map. Now look at the concept map for Chapter 49 on
page 476. Notice that the concept map has been started for you. Add the key
facts and concepts you feel are important for Section 49–4. When you have
finished the chapter, you will have a completed concept map.

## Using the Writing Process

Use your writing skills and imagination to respond to the following writing assignment. You will probably need an additional sheet of paper to complete your response.

Many people feel that a modification of our lifestyle is necessary in order for future generations to enjoy planet Earth as we do. However, we don't enjoy giving up comforts we have grown used to. Design a campaign that you think would increase the willingness of people to make the necessary sacrifices for future generations.

_____

_____

_____

_____

_____

_____

_____

_____

_____

_____

_____

_____

_____

_____

_____

_____

_____

_____

_____

_____

_____

## Concept Mapping

The concept map below has been started for you. Add the key facts and concepts for each section of the chapter to this partial concept map. When you are done, you will have a concept map for the entire chapter.

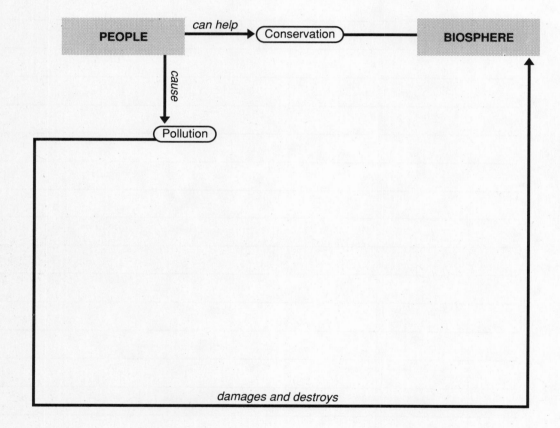